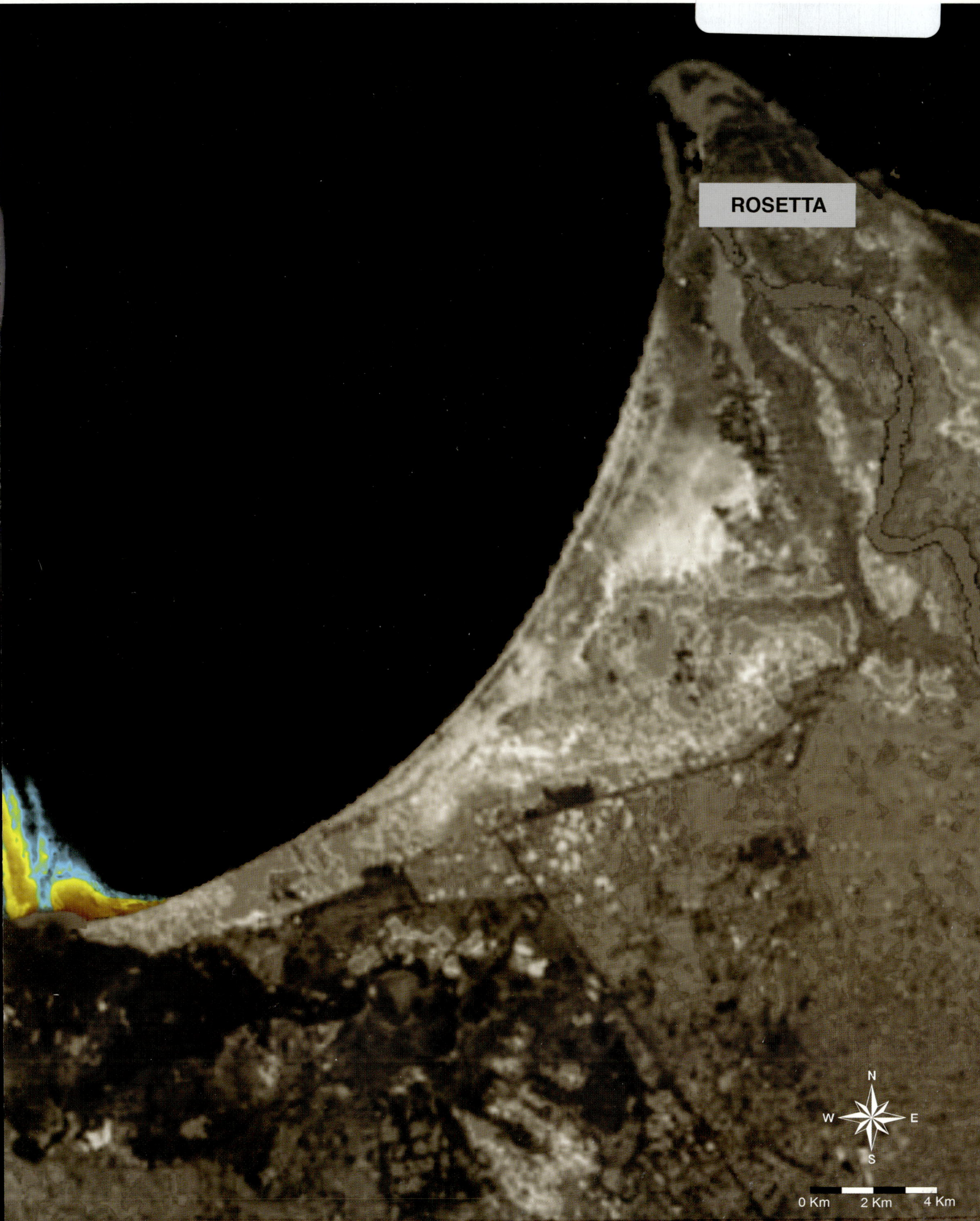
ROSETTA
N
W
E
S
0 Km
2 Km
4 Km

Underwater Archaeology in the Canopic Region in Egypt

Geoarchaeology

Jean-Daniel Stanley

with

Alessio Bandelli, Maria Pia Bernasconi, Thomas Jorstad, Romana Melis, Nevio Pugliese, Gérard Schnepp and Andrew G. Warne

Series editor: Jonathan Cole

Oxford Centre for Maritime Archaeology: Monograph 2
Institute of Archaeology, University of Oxford
2007

Published by the
Oxford Centre for Maritime Archaeology,
Institute of Archaeology,
36 Beaumont Street, Oxford OX1 2PG

ISBN 978-0-9549627-4-6

HILTI FOUNDATION

The missions of the European Institute for Under Water Archaeology in Egypt have been supported since 1996 by the Hilti Foundation and have been undertaken in cooperation with the Supreme Council of Antiquities, Egypt, under the General Secretary Prof. Dr. Zahi Hawass.

Designed and produced by Charlie Webster
Printed in Great Britain at the Alden Group, Oxfordshire

List of Contributors

Alessio Bandelli
Dipartimento di Scienze Geologiche, Ambientali e Marine, Università di Trieste, Via Edoardo Weiss 2, 1-34127 Trieste Italy

Maria Pia Bernasconi
Dipartimento di Scienze della Terra, Università della Calabria, 1-87036 Rende (CS), Italy

Thomas F. Jorstad
Geoarchaeology Program, E-205 NMNH, Smithsonian Institution, Washington, DC 20013-7012, USA

Romana Melis
Dipartimento di Scienze Geologiche, Ambientali e Marine, Università di Trieste, Via Edoardo Weiss 2, 1-34127 Trieste Italy

Nevio Pugliese
Dipartimento di Scienze Geologiche, Ambientali e Marine, Università di Trieste, Via Edoardo Weiss 2, 1-34127 Trieste Italy

Gérard Schnepp
Institut Européen d'Archéologie Sous-Marine, 123 Route de Saint Nizier, 38170 Seyssinet-Pariset, France

Jean-Daniel Stanley
Geoarchaeology Program, E-205 NMNH, Smithsonian Institution, Washington, DC 20013-7012, USA

Andrew G. Warne[†]
Geoarchaeology-Global Change Program, E-205 NMNH, Smithsonian Institution, Washington, DC 20013-7012, USA; and US Geological Survey, 651 Federal Drive, Guaynabo 00965, Puerto Rico
(† deceased)

Contents

List of Figures

Chapter 2

Chapter 3

Chapter 4

Chapter 5

List of Tables

Preface

This book – the second in the Oxford Centre for Maritime Archaeology series on *Underwater Archaeology in the Canopic Region in Egypt* – marks a major step forward in our understanding of the north-west Nile Delta in antiquity. It follows the general topographic study by Franck Goddio in *The Excavation and Topography of Heracleion-Thonis and East Canopus*, and will be followed in turn by a number of detailed studies of specific monuments and archaeological material.

The geoarchaeological investigation of the Aboukir Bay region, undertaken in concert with the underwater excavations by the Institut Européen d'Archéologie Sous-Marine (IEASM) with the financial support of the Hilti Foundation, demonstrates the importance of innovative and interdisciplinary studies in maritime archaeology, especially in the investigation of submerged land sites.

Aboukir Bay, where the Battle of the Nile was fought in 1798, did not exist in antiquity. Instead, the region was a promontory sheltering the Canopic mouth of the Nile, a complex landscape of river branches and canals, sanctuaries and settlements. Foremost among these was the lost site of Heracleion-Thonis, the main trading harbour of Egypt before the foundation of Alexandria, with a major temple. To the west lay another settlement which later became the site of an important Byzantine monastery. By the early Arab period, this whole landscape, with its sites and sanctuaries, both pagan and Christian, had sunk beneath the waves. This book explains how, and why.

The submergence of the sites of East Canopus and Heracleion is partly due to relative sea-level changes as a result of eustatic sea-level rise and subsidence of the North African landmass. But these factors cannot account for the full depth of subsidence; the remainder is shown to be a result of the failure of unconsolidated deltaic sediments when entrained moisture is suddenly expelled as a result of increased loading, under conditions such as might occur during an earthquake, tsunami or an exceptionally high Nile flood. The geomorphological and archaeological record may suggest several such events, including one in the mid eighth century AD.

The work presented here by Jean-Daniel Stanley and his collaborators provides a firm foundation for a new understanding of the Canopic region in antiquity, and is essential for providing a topographic interpretative framework for understanding the archaeological remains discovered by the IEASM missions. This detailed study is not merely of value for the historical topography of Aboukir Bay and the environs of ancient Alexandria, but also as a study of the historical geomorphology of the region and the evolution of deltaic environments. The School of Archaeology in the University of Oxford, through the Oxford Centre for Maritime Archaeology, is proud to be associated with the IEASM and the Hilti Foundation in the publication of this innovative programme of underwater fieldwork.

Andrew Wilson
Professor of the Archaeology of the Roman Empire
School of Archaeology, Oxford University

Acknowledgements

Chapter 2

Mr Gérard Schnepp at Arcocea International, Grenoble, France, provided the detailed bathymetric maps of Aboukir Bay which he compiled and which are presented in Figure 2.12. Mr Thomas Jorstad, Ms Mary Parrish and Ms Jaime Siebrase at the National Museum of Natural History, Smithsonian Institution, Washington DC, provided helpful technical assistance in the preparation of this chapter. Dr Robert H. Meade (US Geological Survey) pointed out key reference sources. We express our appreciation for research grants (to J.-D.S.) that enabled us to undertake the present study: from the National Museum's Walcott Fund, and from Franck Goddio, President of the Institut Européen d'Archéologie Sous-Marine, Paris. We thank Dr Robert Hirsch, Director of the Water Resources Division, US Geological Survey, for providing funds (to A.G.W.) to research and write this chapter.

Chapter 3

Franck Goddio, the Institut Européen d'Archéologie Sous-Marine (IEASM) and the Hilti Foundation gave vital support and field data needed to conduct this study at the National Museum of Natural History (NMNH), Smithsonian Institution, Washington, DC. During the course of the investigation, we have been assisted by numerous specialists, and sincerely thank all of them, including: J.-P. Blancan (IEASM), J.-J. Groussard (ARCOCEA) and D. Visnikard (IEASM), for bathymetric, magnetometer, side-scan, seismic and other surveys at sea; J.-C. Aussa (Triton Electronics International), for the sub-bottom seismic survey; Drs A. Rebora and S. Laura and the geological exploration staff at IEOC-AGIP in Cairo, for seismic information on Aboukir Bay; F. Pereira (IEASM), for field photographs; K. Hayes and J. Nickels (Aqua Surveys Inc.) and E. Smith (IEASM), for vibracoring; Drs M.P. Bernasconi (University of Cosenza), R. Mellis, and N. Pugliese (University of Trieste), for analyses of the faunal content of surficial samples at East Canopus; Dr A.G. Warne (US Geological Survey, deceased) and K. Baker, E. Berman, N. Cox and Y. Crespo (interns at the NMNH), for help with laboratory analyses of vibracore sediment; and M. Parrish (NMNH), for help with illustrations. We also very much appreciate reviews of this chapter by Drs W.G. Melson and M.R. Senatore.

Chapter 4

Messrs K. Hayes and J. Nickels, Aqua Survey, Inc., and E. Smith, Key West, Florida, assisted with recovery of vibracores in Aboukir Bay, and Mr T.F. Jorstad, Smithsonian's National Museum of Natural History (NMNH), helped with laboratory analyses and reviewed the manuscript. Support for the fieldwork and funding of laboratory research was provided by Mr Franck Goddio, President, Institut Européen d'Archéologie Sous-Marine (IEASM), Paris, the Walcott Fund at NMNH, and a grant from the Marine Initiative Program at the Smithsonian Institution.

Chapter 5

The Institut Européen d'Archéologie Sous-Marine (IEASM) encouraged and funded this project. Many members of the IEASM team assisted in the coring operation, especially Eric Smith. Numerous collaborators, both inside the Smithsonian Institution and in other agencies, businesses and universities, provided their support and expertise towards this investigation. At the Smithsonian's NMNH they include: Fran Hueber, Kate Baker, Elise Berman, Yorka Crespo, Nicole Cox, William Boykins, Don Dean and Elizabeth Valulis of the Department of Paleobiology; Carolyn Rose (deceased), Agnes Stix, Sarah Pelot, Doug Owsley, Gus van Beek, Don Ortner, Steve Ousley, Bruno Frolich and Bruce Smith of the Department of Anthropology; Robert Faden, Robert Soreng, Paul Peterson, Dan Nicolson and Aaron Goldstein of the Department of Botany; and Scott Whittaker and Susann Braden of the Scanning Electron Microscopy Laboratory. The following specialists assisted at the Center for Materials and Education Research at the Smithsonian's Museum Support Center: Harry Alden, Melanie Feather and Ron Cunningham. Colleagues in other organisations include: Andrew Warne (deceased) of the US Geological Survey; Byron Burk, independent contractor; Alex Kowal and Mike Holloway of Eastman Kodak Company; Ken Hayes and Jim Nickels of Aqua Survey Inc.; Wilma Wetterstrom and Deena Decker-Walters of Harvard University; Ed Reinhardt of Macmaster University, Canada; Janus Dominik of the University of Geneva, Switzerland; and Fabrice Monna of the University of Bourgogne, France.

Abbreviations

AMS	Accelerator Mass Spectrometry
CT	Computed Tomography
GPS	Global Positioning System
IEASM	Institut Européen d'Archaéologie Sous-Marine
NMNH	National Museum of Natural History (Smithsonian Institution)
OCMA	Oxford Centre for Maritime Archaeology
RMN	Nucleur Resonance Magnetometry
UNDP	United Nations Development Program

Chapter 1

Geoarchaeological Exploration of Submerged Sites in Aboukir Bay, Egypt: an Introduction

Jean-Daniel Stanley

1 Geology integrated with submarine archaeology

Geoarchaeology is the integration of earth sciences with history and archaeology to interpret past natural conditions more precisely and comprehensively, and also to identify the effects of anthropogenic activity in areas of earlier human occupation. It should be recognised that geologists and archaeologists have key common interests as regards site exploration. Of significance to both are the critical elements of time and the ability to reconstruct, in three dimensions, past sedimentary and anthropogenically modified stratigraphic horizons. Thus, to interpret coherently the history of an early settlement, and especially one positioned underwater, both geologists and archaeologists need to coordinate their efforts closely, so as to place a former site within as precise a time-frame as possible and define the site's environmental setting before, during and following human occupation.

To envision how one would interpret the history of a site now submerged in Aboukir Bay, off Egypt's north-west Nile delta, it might be useful to consider the approach one could take to study, say, a modern city such as Venice, New Orleans or any other vulnerable low-lying coastal centre prone to subsidence and potentially serious damage. What would present-day scholars need to record in detail about such a modern city if they foresaw that it might one day disappear beneath the waves? In such a case, Charles Lyell's principle of 'the present as the key to the past' comes to mind and almost certainly would be a powerful approach for the merging of submarine geological and archaeological exploration efforts. One would inevitably focus on describing and measuring prominent features before such a city's disappearance, and especially those that could provide valuable historical information for scholars at a later date. Thus, application of Lyell's principle, as well as its reverse (the past as key to the present), is the basis of our approach and underlies the manner in which investigations were conducted and summarised in the following four chapters prepared for this monograph.

Any comprehensive study, whether by geologist or archaeologist, of a now submerged site of Greek to Byzantine age in Aboukir Bay must, at the very least, attempt to resolve the following key questions: (1) Where were the sites originally positioned relative to the Egyptian coast and the Nile branch between the sixth century BC and the eighth century AD, and why were they placed in such a location? (2) What was the nature of the geography and environmental conditions of the site localities while subaerially exposed, and how did these change through time? (3) What processes, natural and anthropogenic, were responsible for their destruction and subsidence beneath the waves?

The geoarchaeological technology necessary to resolve these questions in the study area includes: geographical analyses of the bay-floor and once contiguous land areas, by means of closely spaced bathymetric surveys; geological study, primarily by means of petrological, biofacies and geochemical investigations of closely spaced radiocarbon-dated core sections; and geophysical surveys using tight-grid high-resolution sub-bottom seismic, side-scan sonar, and nuclear resonance magnetometer technologies.

We have found it necessary, where possible, to proceed with the following steps to study Heracleion and East Canopus, now submerged in the western part of the bay:

- Dating and describing core samples and any artefacts therein, to determine when and where the sites were initially emplaced.

- Detailing the geography of each settlement and its surroundings prior to submergence, focusing on the overall environmental setting. This comprehensive analysis would include identification of the following features and processes: the original coastline, river mouth and channels; elevation of the settlement above sea level; coastal oceanographic parameters (tides, winds, currents and waves); normal and exceptional climatic conditions; and petrological attributes and engineering properties of the sediment substrate on which the settlement was positioned.

- Identifying attributes of normal environmental conditions and distinguishing these from much stronger episodic processes. These latter include high waves driven by winter storms, and the coincident interaction of high tides with strong winter storm surges. Some of these less frequent, but powerful, events probably impacted the coast and the sediment substrate on which near-shore settlements were placed.

- Recognising even less frequent but more powerful natural events (Nile floods, earthquakes, tsunamis) and also anthropogenic effects that could have induced destruction of a vulnerable low-elevation site. Did construction practices and emplacement of large structures without foundations on an unconsolidated sediment base lead directly, or indirectly, to eventual failure of such a site? What protective measures did populations implement (if any) to protect their settlement, and to what extent were they effective?

- Measuring the date, and how long a period was involved, for the complete submergence of the two sites in Aboukir Bay.

Initiation of the research presented here was achieved by collaboration between the IEASM exploration team and specialists affiliated with the Smithsonian Institution's geoarchaeology programme. Appreciation is expressed to Mr Franck Goddio, the Institut Européen d'Archéologie Sous-Marine (IEASM) in Paris and the Hilti Foundation for their encouragement and support of our investigations.

2 Among the major findings

The chapters in this volume focus on defining the former geography, identifying the effects of natural processes and human activity, and recording the sudden changes of settlement substrate as measured by evolution of fauna and biofacies at the two sites in the study area. Our research to date provides information on a host of different topics, several of which are briefly summarised in text sections below.

3 Palaeogeographic interpretation of settlement areas

Interpretation of the diverse environments where sites were positioned is based on recently acquired bathymetric, sediment core and geophysical data. Detailed lithological logs for the 21 vibracores collected at the two sites, and areas between them, have been compiled. Among the features recorded in core sections are: lithology, colour, sedimentary structures and markings made by organisms in the sediment, positions of radiocarbon-dated samples, organic-rich layers such as peat, rock fragments and human artefacts. The latter include potsherds and rock material (lithoclasts) of Greek, Roman and Byzantine to modern origin. In addition, radiocarbon dates, both conventional and AMS, were successfully obtained for 58 core samples to establish coherent time-stratigraphic baselines.

Aboukir Bay was formed by subsidence of the promontory of the Nile's Canopic branch and development of the younger Rosetta branch and its promontory to the east. Prior to submergence, the Canopic branch had formed a large promontory (area ~55 km^2) in what is now the western sector of Aboukir Bay. The promontory extended well into the Mediterranean (~14 km to the north of the present coast), and much of the Nile's quartz-rich sand was transported northward and to the northeast of this feature. Coastal settlements were built, where possible, on somewhat higher mounds of sand near the coast, which, at the time, rose several metres above sea level. Marked coastal changes that occurred during the period of site occupancy were largely the result of interaction of progressive sea level rise and land subsidence, and concurrent nearshore sediment erosion by waves and coastal currents.

Our estimates of sea level rise in this sector take into account the role of world-wide (eustatic) rise plus local effects of land lowering by sediment compaction and isostatic depression. Analyses indicate that relative sea level rise in this region has ranged to >1.0 mm/yr, or more than about 2.5 m during the past 2500 years. Archaeological findings also show that the sites totally subsided at substantially the same period of time, i.e. the second part of the eighth century AD.

Effects of changing coastal physiography, large Nile floods and neotectonics periodically induced important shifts in geographical position of the Canopic channel mouth. Studies by previous authors have suggested that major flow in the Canopic Nile channel ended prior to about the third–fifth centuries BC. Our surveys, on the other hand, record that some flow probably lasted until the tenth century BC. Even after that time, Nile water, albeit much diminished, continued to pass across this north-west sector of the Nile delta margin, as recorded by mapping of relict traces of the Canopic channel.

3.1 Subsidence mechanisms

Settlements, originally positioned at, or just landward of, the coasts were particularly vulnerable since they were built on a soft, physically unstable, water-saturated sediment substrate lying just a few metres above sea level. The molluscan and microfossil (foraminifera, ostracod) fauna in core sections serve as sensitive markers of environmental changes at site localities, and provide useful information on subsidence of sediment substrates upon which the two ancient sites were originally built. Distinct faunal variations in the dated cores help define the timing of subsidence. Poorly preserved molluscan fauna associated with iron-stained microfossils denote sudden lowering of brackish lagoonal sediment settings beneath fully saline marine water. Such damaging events occurred earlier at Heracleion than at East Canopus. Both core lithology and faunal investigations record lowering caused by long-term gradual and progressive processes (sea level rise, land submergence) coupled with a series of episodic sudden and destructive events (such as Nile floods, storm surges, earthquakes and tsunamis).

Integration of the faunal record in cores with geological evidence (disturbed and slumped bedding, evidence of liquefaction phenomena in cores) and geophysical survey data (diapirs, faults, offset strata on seismic profiles) provides powerful evidence for abrupt subsidence events that led to disappearance of the two settlements. Moreover, engineering properties of the rapidly deposited, water-saturated Nile sediment that had accumulated at and near the mouth of the Canopic branch indicate that site substrates were prone to failure. Our measurements indicate that gradual eustatic rise and depression of land alone would have accounted only for submergence of less than 3 m, or less than half of total site lowering depths on the bay-floor at which they now lie. The result of gradual submergence processes plus sudden sediment failure resulted in a total lowering of sites to 5–7 m below present mean sea level.

Sedimentary structures, such as disturbed and offset bedding, in the dated core sections indicate that high Nile floods were among the important, and perhaps most frequent, triggers of sediment failure. This is substantiated by examination of Nilometer river level data collected for that time period. For example, archaeological artefacts such as the most recent coins recovered at East Canopus are dated before the particularly high flood documented in 741 (or 742) AD. However, this is not the case for Heracleion where an Islamic coin (785 AD) was found in a context of other artefacts of the 7th–8th century AD.

Although the importance of high Nile floods is emphasised, it is recognised that sudden loading and triggering of sediment failure by storm surges, earthquake tremors and tsunamis also occurred periodically. It is probable that some substrate failure was also caused by weighting (loading) effects resulting from human activity. For example, emplacement of large and heavy walls and temples directly on a soft sediment substrate probably accounted for some collapse. It is of note that comparable mass-failure phenomena in Alexandria's western harbour were recorded by engineers less than a century ago as a result of triggering by port construction projects. This activity, leading to sudden substrate subsidence of as much as 3–5 m, appears to be a useful modern analogue with which to interpret some past anthropogenic failure events.

4 Questions remain, studies continue

The findings obtained by earth scientists provide new baselines and a preliminary geoarchaeological framework to assist marine archaeologists refine the history of sites in Aboukir Bay. Yet it is recognised that these contributions are but a beginning. Listed here are some topics that warrant additional consideration, require further exploration and analysis, and are the object of on-going research.

Transformation of the once subaerially exposed Canopic delta platform into a bay-floor surface.

More precisely than now measured, when, why, by how much and how fast did this transformation to a bay occur? To what extent can submergence and bay-floor formation be related to geologically recent tectonic shifts of the Pleistocene coastal limestone ridge (kurkar) exposed at Aboukir (ancient Canopus) and which, presently, underlies Holocene sediment in the bay proper? Did shifts of this carbonate cemented substrate involve stretching, down-bowing and offset of the basement in a north-east direction? Or alternatively, has lowering of the bay-floor been caused by shifts and structural readjustments at much greater depths within the thick (>5000 m) sequence of underlying sedimentary strata?

Role of earthquakes.

Have shifts and readjustments within the thick underlying pre-Pleistocene sedimentary sequence been responsible for episodic intermediate and shallow-depth earthquakes in proximity of the bay?

To what extent were some of these neotectonic events responsible for damage to the now submerged sites during the past three millennia?

Significance of Nile floods.

In what manner and how frequently did unusually high Nile floods cause damage to East Canopus and Heracleion, which once were positioned near the mouth of the Nile's Canopic branch? What are the criteria, geological, biological and anthropogenic, that can be used reliably to distinguish flood from earthquake and tsunami damage?

Better definition of the Nile's Canopic branch and potential discovery of new submerged sites.

To what extent did offset and submergence of the lower Nile delta margin cause lateral shifts of the Canopic channel and submergence of settlements in Aboukir Bay? Precisely, until how recently (early, mid or late first millennium BC) was the Canopic a truly active fluvial channel? Were other established sites abandoned as the Canopic channel migrated eastward across the delta plain during the first millennium BC? If so, where would these yet-to-be-discovered settlements probably be positioned relative to the modern coastline and thick sediment accumulations west of the Damietta branch and its promontory?

New methods and technology to define better the bay-floor and abandoned Canopic branch on the lower Nile delta plain.

For example, can mapping by Shuttle Radar Topography Mission (SRTM) and other satellite systems serve to discover new sites such as those that may have been abandoned when the Canopic channel shifted periodically and migrated laterally across the delta plain? Can traces of ancient canals (some perhaps pre-Ptolemaic) be identified and delineated? To what extent can geochemistry be applied to detect early human activity in the study area, including evidence of palaeopollution and effects of metallurgy?

Applying findings obtained in Aboukir Bay to initiate exploration and mapping of other ancient sites along the Nile delta margin.

To what extent are the timing and processes responsible for submergence and burial in the bay applicable for ongoing and new exploration of other ancient settlements elsewhere along the delta coast? Would such findings serve better to investigate submerged sites in the Alexandria harbours and at Tel Tennis and Pelusium? What new information can be obtained on pre-Ptolemaic activity, such as the early part of the Late Period in Alexandria's eastern harbour, and even older settlement evidence found at the Mediba site to the east, near the coast of Burullus lagoon?

The merging of earth science subdisciplines with coastal and submarine archaeological exploratory techniques, a relatively new approach, is proving to be a productive endeavour. It is anticipated that continued involvement of collaborative geoarchaeological projects will help further clarify some key unanswered questions pertaining to the sites in Aboukir Bay.

Chapter 2

Nile Delta Geography at the Time of Heracleion and East Canopus

Jean-Daniel Stanley and Andrew G. Warne

'. . . he came to Egypt . . . to what is now called the Canopic Mouth of the Nile and to the Saltpans. There was, on the shore (it is still there), a shrine of Heracles.' Herodotus, *The Histories*, 2.11

In Herodotus' time, a seaman steering his ship towards the mouth of the Canopic branch, then the largest of the Nile distributaries, observed a very different coastal configuration than that seen today. In the fifth century BC, ships did not sail into what is presently called Aboukir Bay, but rather headed towards a large coastal promontory associated with the former Canopic channel of the Nile.

In this synopsis, we present geographical data pertaining to the north-west Nile delta margin and its prevailing river system to understand better why, where and how the Egyptians, Greeks, Romans and their descendants positioned the coastal sites of Heracleion and East Canopus between the sixth century BC and the eighth century AD. We present evidence that western Aboukir Bay did not exist at the end of the Dynastic or even the Ptolemaic period, but formed after the beginning of the first millennium AD during the Roman and Byzantine periods. Information on major fluvial and near-shore processes is integrated with radiocarbon-dated core data to develop a comprehensive palaeogeographical framework for the north-west Nile delta during the time of Heracleion and East Canopus.

1 The Nile delta setting

The Nile delta has formed at the mouth of one of the world's longest rivers (6690 km) that drains a ~2,880,000 km^2 basin. The delta plain encompasses about 22,000 km^2 and is delineated to the north by a wave-cut, arcuate coast that is ~225 km long between its western margin near Alexandria and its eastern boundary in the western Sinai (Figure 2.1). The delta radius, from its apex at Cairo to the coast, is ~160 km, and the delta plain elevation decreases from ~18 m at Cairo to <1 m along the coast. Numerous active distributary channels radiated northward from the delta apex at Cairo (Toussoun 1922, 1926, 1934; Ball 1939, 1942; Stanley *et al.* 1993). The historic Nile delta plain was composed of relatively well-drained arable lands in the upper and middle delta, and extensive, poorly drained wetlands in the lower delta (Figure 2.1). Laterally continuous sandy beaches and dune fields, together with silt-rich promontories at the mouths of major distributary channels, comprised the northern coastal sector (Manohar 1981; Smith and Abdel-Kader 1988; Frihy and Komar 1991; Nafaa and Frihy 1993).

Most river Nile sediments are derived from the Blue Nile and Atbara river basins whose head-waters are in the Ethiopian Plateau of east-central Africa. Summer monsoon rains produce pronounced seasonal flow in the Blue Nile and Atbara rivers, which in turn causes the annual Nile flood (Hurst 1931–66; Butzer and Hansen 1968; Said 1981; Shahin 1985). The more continuously flowing White Nile, which drains the larger Central African (Kenyan) Plateau, carries much less sediment to the lower Nile valley, because a large proportion of its sediment is deposited in the Sudd swamps and marshes of southern Sudan (Hurst 1931–66; Shukri 1950; Hassan 1981). The Nile basin north of the Sudd is hyper-arid, and very little additional water flows into the river Nile north of the Atbara tributary.

The Nile has generally low discharge in comparison to many shorter rivers with smaller drainage basins (Coleman 1982). Annual average discharge exceeded 100 billion m^3 in the nineteenth century (Sestini 1992a,b; Said 1993), of which more than 80% of Nile discharge occurred from August to October (Figure 2.2, right inset). Before emplacement of barrages and dams in the Nile valley and delta, approximately 93–98% of the sediment load reached the delta plain and coast, with much of this material discharged into the Mediterranean Sea during peak flood months of August and September (Holeman 1968; El Din 1977; Waterbury 1979). Floods at that time carried a sediment load composed of about 25% sand, 45% silt and 30% clay to the delta (El Din 1977).

Marine waves and currents partially remove and extensively displace river sediment transported to the coast; overall, however, the north-west delta has prograded seaward during the Holocene (Stanley and Warne 1993). During the past seven millennia, silt accreted within the lower delta plain at average rates of roughly 1–7 mm per year (Stanley and Warne 1993a,b, 1998). The thickness of Holocene strata tends to increase northward towards the coast, where it ranges from 10 to 50 m beneath the lower delta plain.

Annual Nile floods (Figure 2.2, right inset) have been a key element in the evolution of the delta and in the development of Egyptian civilisation. In a typical flood season, large portions of the delta plain were inundated and uninhabitable but most of the delta plain drained within a couple of months as water level declined as a response to evaporation, soil infiltration and lowered groundwater level (Butzer 1976). The Nile valley and delta, however, have been subject to unpredictable fluctuations of Nile flow (Popper 1951; Hassan 1981; Said 1993). Moreover, annual variation of annual peak Nile flood stage by as little as 1 m could mean the difference between famine (flood stage too low) and widespread lower delta and coastal destruction (flood stage too high). In certain periods, such as the end of the Old Kingdom and First Intermediate Period (about 2250–1950 BC), unusually low annual discharge and flood levels were a major impetus to develop water-management projects in order to avert famine. During early cultural phases, Egyptians modified natural flood basins on the delta plain to entrap flood water and nutrient-rich sediment. Despite these modifications to the delta plain, major floods were highly destructive to sites located along major distributary channels (De Cosson 1935).

2 The recent north-western delta

The Aboukir shoreline is concave, in contrast to the convex coastline of the central and eastern Nile delta (Figure 2.1). The Aboukir coast extends about 50 km from the Aboukir carbonate ridge in the west of the bay to the siliceous sand and silt-rich Rosetta

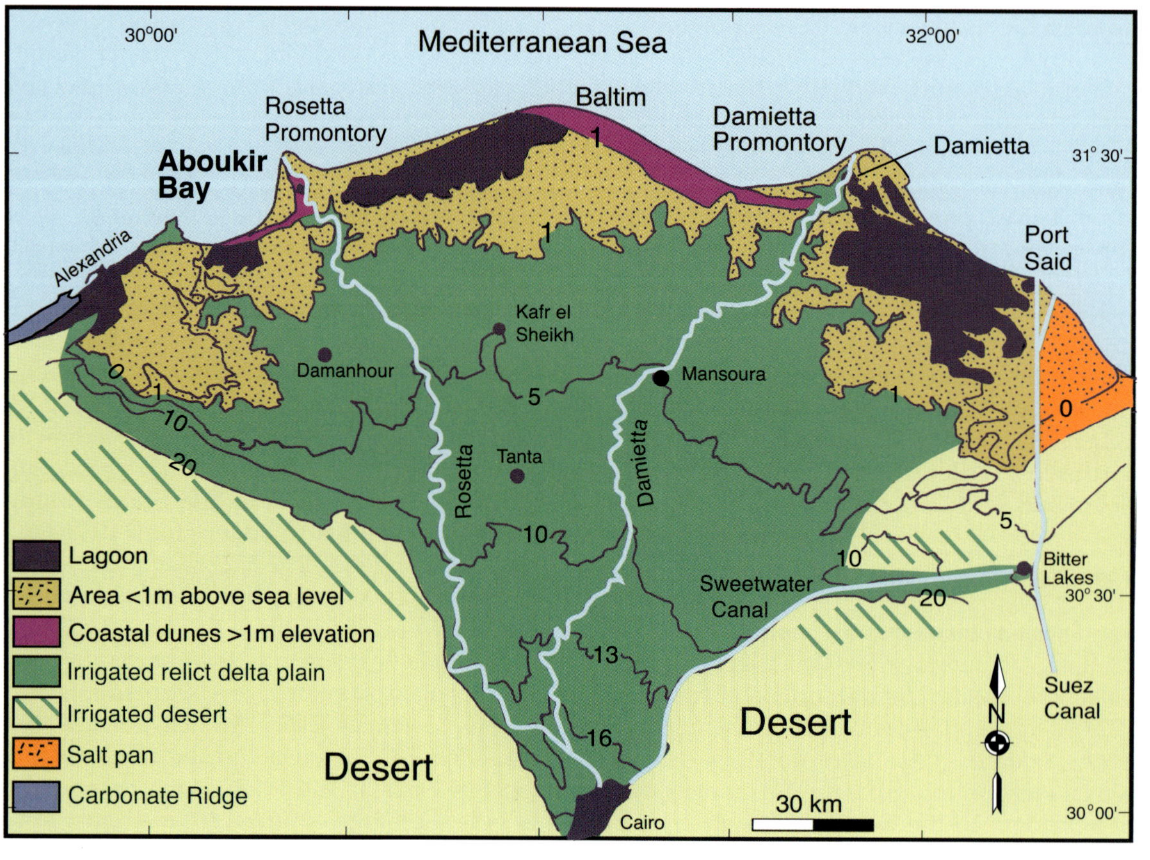

Figure 2.1 Principal topographical and geographical features of the Nile delta, Egypt. Elevation contours are in metres. Modified from Sestini (1992a,b).

promontory in the east (El Askary and Frihy 1986) (Figure 2.3A). The Aboukir ridge, formed of consolidated upper Pleistocene limestone, or kurkar, extends from the modern town of Aboukir north-eastward about 5 km into the Mediterranean. On the eastern side of the bay, the triangular mass of unconsolidated sediments that comprise the Rosetta promontory protrudes seaward towards the north-north-west. In addition to Aboukir ridge and Rosetta promontory, the modern bay region comprises several distinct environments (Figure 2.4): beach, backshore sandflat and strand plain, coastal dune, brackish to freshwater wetland (lagoon, marsh), kôm (low hill), drain and canal, and agricultural land to the south (Figure 2.4) (El-Fayoumy *et al.* 1975; El Fishawi and El Askary 1981; Frihy *et al.* 1988; Chen *et al.* 1992; Stanley *et al.* 1992; Stanley and Hamza 1992; Warne and Stanley 1993).

The coastal region of the modern bay can be subdivided into two distinct sectors on the basis of shoreline configuration and near-shore bathymetry: the Aboukir ridge to El Maadiya sector in the western bay is irregular, whereas the longer El Maadiya to Rosetta sector to the east is smooth (Figure 2.4). Moreover, bathymetric contours in the eastern bay record an even, gentle near-shore gradient, in contrast to those of the west. The sea-floor on which the East Canopus and Heracleion sites are located in the western part of the bay is a triangular to crescentic-shaped, shallow platform (for the most part <7 m deep) that extends northward about 12 km from the southern bay shore (Figure 2.3A) (UNDP/UNESCO 1978). Differential GPS was used to locate the centre of East Canopus (N 31° 18′ 880, E 30° 5′ 350) and Heracleion (N 31° 18′ 775, E 30° 7′ 670).

A mostly submerged portion of Aboukir ridge extends seaward (north-eastward) at shallow depths; the submerged carbonate ridge physically separates the north-west bay from the open Egyptian Shelf. The site of the now, Nelson Island, about 4 km north-east of Aboukir, is located along this carbonate ridge system (Figure 2.3A). At the present town of Aboukir, the ridge rises to about 10 m above sea level; the ridge extends south-westward through Alexandria, and forms the linear Egyptian coast to Arab's Gulf (Butzer 1960; Alexandersson 1990; Stanley and Hamza 1992; Hassouba 1995).

Beach and shallow marine sediments along the western margin of Aboukir Bay contain large proportions of sand-to-pebble size carbonate

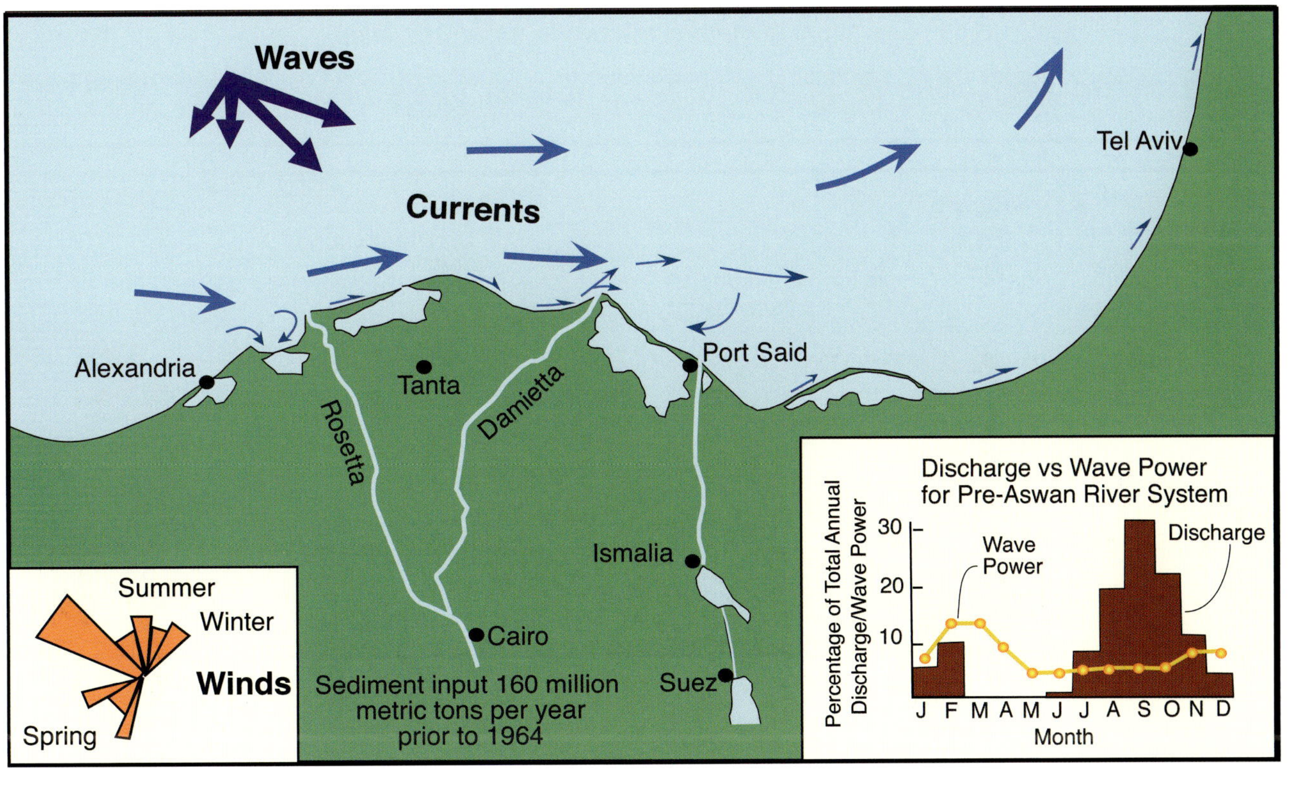

Figure 2.2 Summary of wind, marine wave and coastal current regime of the Nile delta coastal region. Note the strong easterly components of the longshore littoral currents. The inset in the lower right summarises the relationship between sediment discharge of the pre-Aswan Dam river Nile and the waves that rework the sediment. The inset in the lower left summarises the Nile coast wind regime, which is dominated by winds to the south-east.

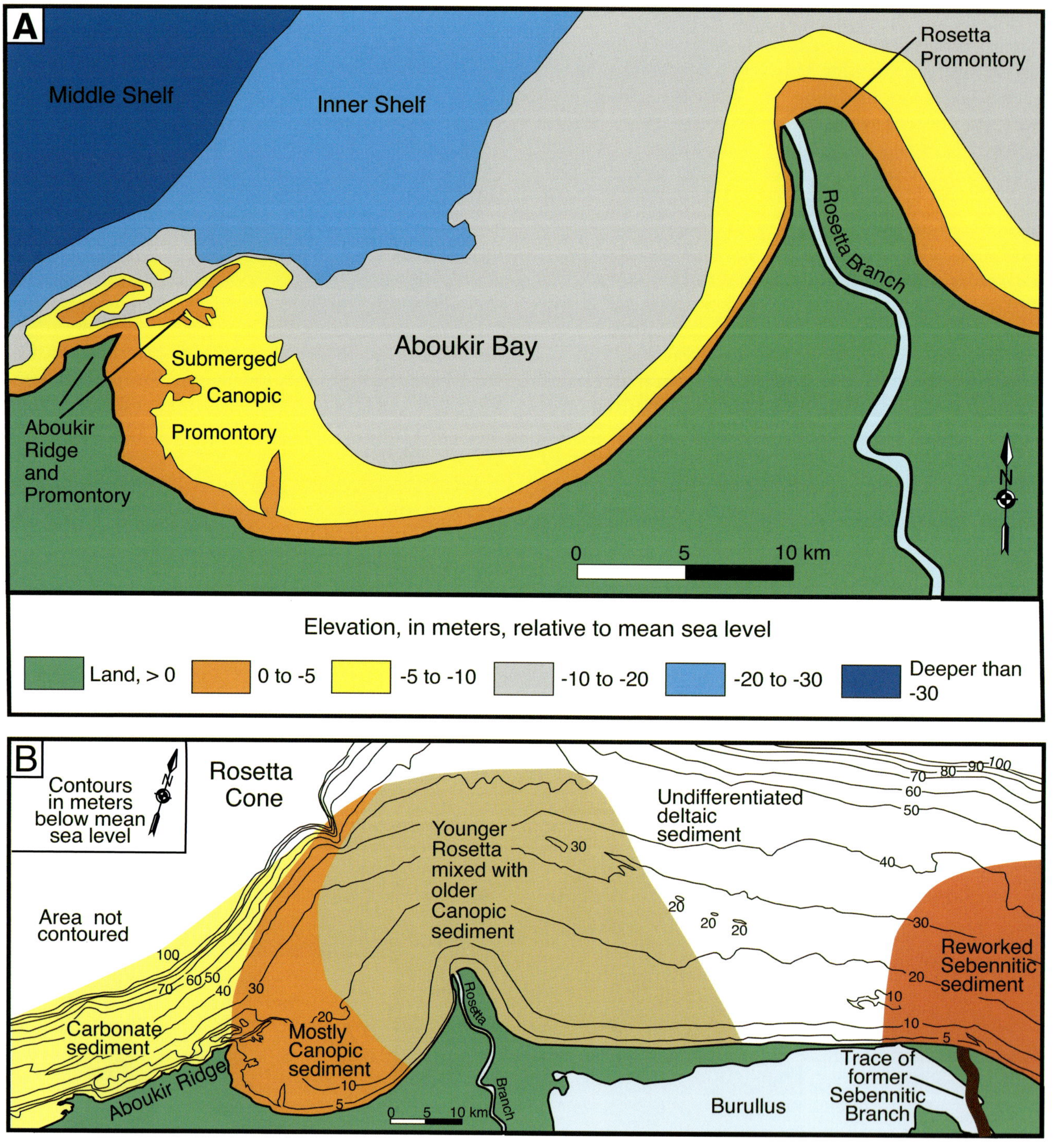

Figure 2.3 (A) General bathymetry in the area from Aboukir to the Rosetta promontory along the north-western Nile delta. Note the submarine ridge that extends north-eastward from the Aboukir carbonate headland. (B) General distribution of submarine Nile delta sediments along the north-west Nile delta coast, shelf and slope. The bathymetric base map is modified from UNDP/UNESCO (1978). Note that the Rosetta Cone extends seaward, well beyond the limits of this map.

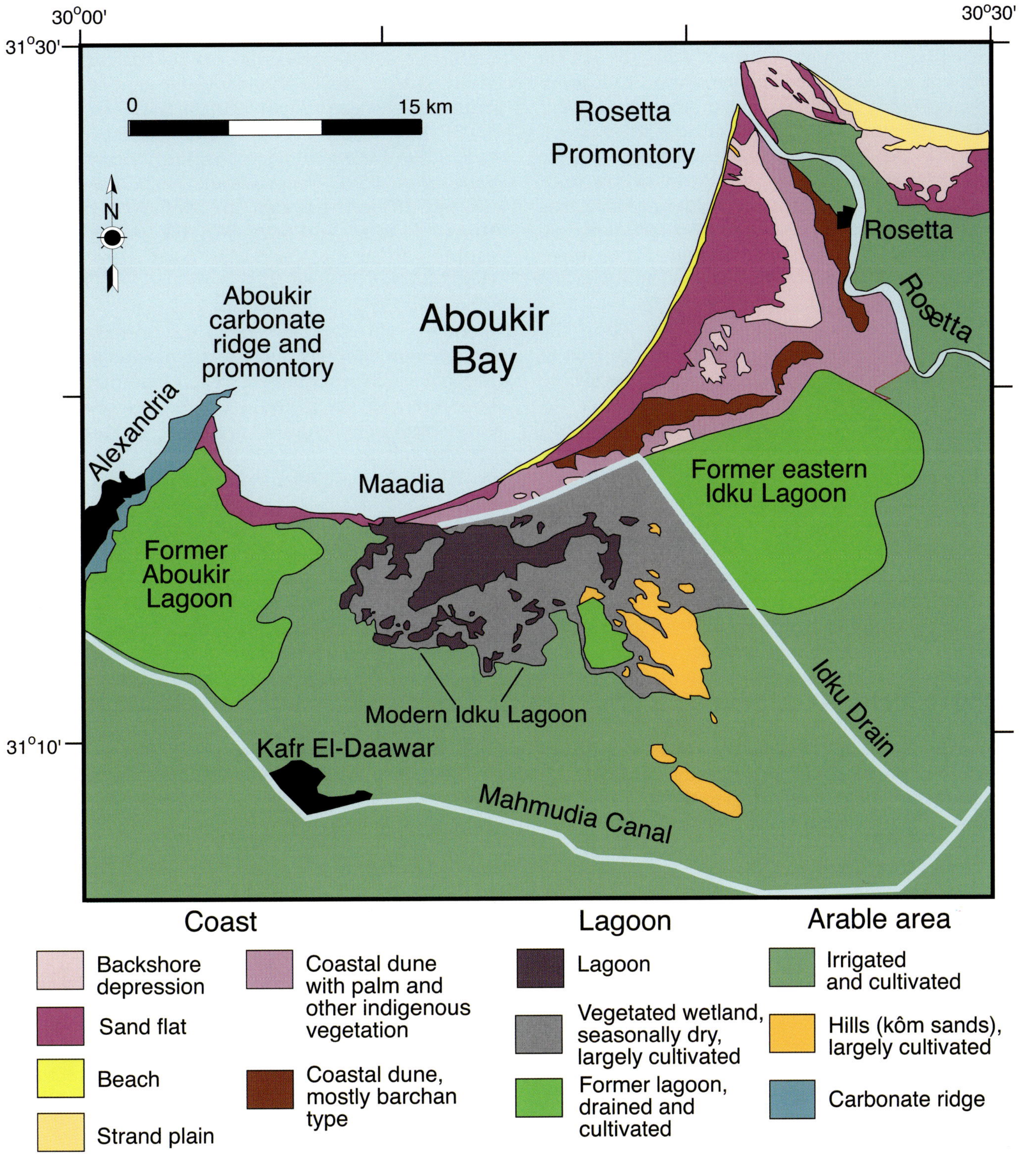

Figure 2.4 Geomorphic/land use map of the modern Nile delta plain in the Aboukir Bay–Rosetta promontory area. This map is based on analysis of Landsat-5 Thematic Mapper satellite imagery taken in December 1986 (IWACO 1989), and on field observations. Modified from Chen *et al*. (1992).

material (Frihy *et al.* 1994). The relative proportion of river Nile (Canopic, Bolbitic-Rosetta) sand, silt and clay increases eastward within the bay (El-Fayoumy *et al.* 1975; El-Wakeel and El Sayed 1978; Summerhayes *et al.* 1978; Stanley and Hamza 1992).

The Rosetta promontory currently extends ~14 km north-north-west into the Mediterranean Sea. Farther seaward, north of the Rosetta promontory, a very large, asymmetric wedge of siliciclastic sediment (Ross and Uchupi 1977) has accumulated along the continental slope (Figure 2.3B). Most of this thick depositional prism, defined as the Rosetta Cone, was formed by sediment discharge from pre-Holocene Nile channels. A substantial portion of the upper section of the Cone comprises current-reworked Canopic branch sediment of Holocene age.

Beaches and backshore sand flats along the Aboukir Bay coastline consist primarily of fine to medium quartz sand of river Nile origin. Backshore sand flats, which are generally <1.0 m above sea level (El-Fayoumy *et al.* 1975), occupy an area of ~55 km^2 and extend to as much as 6 km inland along the north-western flank of the Rosetta promontory. Large coastal dunes, typically 2–16 m in height and 1–3 km wide, are distributed along the bay coast to the west and south-west of the Rosetta promontory (Figure 2.4) (Frihy *et al.* 1988). These dunes were formed primarily by wave reworking of Nile channel sands (El Fishawi and El Askary 1981) and their subsequent transport landward by prevailing north-west winds (Figure 2.2, left inset). Shallow backshore depressions landward of the dunes and along the main channel of the Rosetta promontory extend to depths of ~1 m below sea level; some parts of these depressions are periodically inundated by marine water, particularly during the winter storm season (Loizeau and Stanley 1993). Edkou and formerly Aboukir lagoons were two major brackish water bodies along the north-west Nile delta coast. Aboukir lagoon has been completely drained and is used for agriculture; large portions of Edkou lagoon have also been drained.

3 Processes that shaped the north-west delta margin

The arcuate Nile delta coastline is identified as a wave-dominated system (Galloway 1975). Compared to other coastal regions of the world, the wave regime along the north-west Nile delta is of moderate energy, yet capable of reworking the river sediments discharged to the coast and shelf. The reworked sediments form extensive sandy beaches and strand plains along most of the delta shoreline (Figure 2.1). Development of laterally extensive beach barriers and coastal dune fields separates the shallow wetlands on the delta margin from the open sea (Figures 2.1, 2.4) (Jacotin 1826; Halim and Gerges 1981; Kerambrun 1986; Arbouille and Stanley 1991).

The tidal range in Aboukir Bay is low (spring tides average 30–40 cm), south to south-east directed winds are active most of the year, and a large-scale counter-clockwise littoral current circulation system drives shelf water masses and sediments eastward (Figure 2.2) (Sharaf El Din 1977; UNDP/UNESCO 1977; Coleman *et al.* 1980; Murray *et al.* 1981; Inman and Jenkins 1984; Fanos 1986; Nafas *et al.* 1991; Poem Group 1992). Eastward-directed littoral currents are deflected by the Aboukir ridge and the Rosetta promontory, resulting in counter-clockwise eddy currents in Aboukir Bay (Figure 2.2) (UNDP/UNESCO 1977; Frihy *et al.* 1994).

The present configuration of the north-western delta margin is largely the result of the interaction of river Nile sediment inflow and accumulation along the coast and shelf, and sediment removal and redistribution by marine processes (Figure 2.2). Seaward advance (progradation) of the Nile delta typically prevailed when and where river accumulation rates exceeded the capacity of marine processes to rework the sediment; in contrast, erosional phases occurred when and where coastal processes removed more sediment than river sediment influx (cf. Stanley and Warne 1994).

In addition to the interaction of river and marine processes, other natural factors, including climatic and sea level fluctuations and vertical displacements of the earth's surface, were essential to the development of the north-west delta configuration (Frihy 1988; El Fishawi and Fanos 1989; Stanley 1990; Zaghloul *et al.* 1990; Chen *et al.* 1992; Nafaa and Frihy 1993; Warne and Stanley 1995; El-Sayed 1996). Estimates of global sea level rise during the past 2500 years, excluding regional and local effects of vertical land motion, are typically ~2.5 m (Pirazolli 1987, 1992; Fairbanks 1989). Long-term subsidence rates in the area south-west of Aboukir Bay, based on radiocarbon-dated sediment cores, are low (<1 mm/yr) (Chen *et al.* 1992). However, much higher rates (to ~4.5 mm/yr) are recorded in the Rosetta promontory area; these higher rates are interpreted as a response to rapid deposition, sediment loading and consequent isostatic depression at the Rosetta channel mouth (Chen *et al.* 1992). Moreover, the thick sediment wedge forming the Rosetta Cone north of the Egyptian Shelf (Ross and Uchupi 1977) has probably promoted rapid subsidence in the area.

Nile floods (Hamid 1984) and tectonic processes (Warne and Stanley 1993; Stanley and Warne 1998) may have induced changes in the local configuration of this coastal margin. For example, sudden weighting by flood water at and near the mouths of the Canopic and Bolbitic-Rosetta distributaries may have induced development of growth (listric) faults and localised, rapid subsidence in this coastal region (Stanley *et al.* 2001, and Chapter 3 in this volume). Based on geomorphic, remote sensing and geophysical evidence, several studies have suggested fault displacement in the north-west Nile delta (Shata 1955; Butzer 1960; El Ramly 1968, 1971; El Shazly *et al.* 1975; Hassouba 1995). However, analysis of recorded earthquake epicentres on the Mediterranean margin of Egypt indicates that the north-west Nile delta is largely aseismic (Stanley 1997). This is in marked contrast to areas in the north along the tectonically active Hellenic-Turkish and Cyprus belts, to the north-east in the Levant, and to the south-east in the Gulf of Suez. It is possible, nevertheless, that occasional local tremors and earthquakes and associated tsunamis initiated in more distal sectors of the eastern Mediterranean (De Cosson 1935; Pirazolli 1987; Guidoboni 1994) caused rapid subsidence in Aboukir Bay. For example, an earthquake-triggered tsunami destroyed many structures, and probably caused rapid subsidence, in and around Alexandria's eastern harbour in 365 BC (Guidoboni 1994). Tectonic-induced subsidence would have involved listric faulting, liquefaction and sediment slumping.

4 North-west delta evolution: historical documentation

Nile basin rainfall and, consequently, river Nile discharge have fluctuated markedly during the Holocene (Hurst 1931–66; Popper 1951; Bell 1970, 1975; Quelennec and Kruc 1976; Wendorf *et al.* 1976; Riehl and Meitin 1979; Adamson *et al.* 1980; Hassan 1981; Shahin 1985; Petit-Maire 1989; Said 1993; Stanley and Warne 1998). Analysis of available historic Nile stage records indicates that decadal-to-centennial-scale cycles of wet and dry periods impacted annual river Nile discharge (Popper 1951; Said 1993) (Figure 2.5). These cycles have been a major factor in altering the relative proportion of flow through major Nile distributary channels, and the geographical position of distributary channels and their promontories (Figure 2.6), which in turn altered the configuration of the delta coast through time (Toussoun 1922, 1926, 1934) (Figures 2.7, 2.8).

5 North-west delta evolution from the Greek to the Byzantine period: the geological record

Stratigraphic interpretation of subsurface deposits using radiocarbon-dated sediment core information serves to identify long-term Nile channel and coastal evolution trends in this region where Heracleion and East Canopus were established, flourished and then submerged (Fourtau 1915; Attia 1954; Sestini 1989; Chen *et al.* 1992; Stanley and Warne 1993a,b, 1998; Stanley *et al.* 1996). Subsurface analyses identify numerous Nile distributaries that flowed across the delta plain to the coast, and reveal that the delta coastal configuration changed markedly during the mid-to-late Holocene (Figure 2.9).

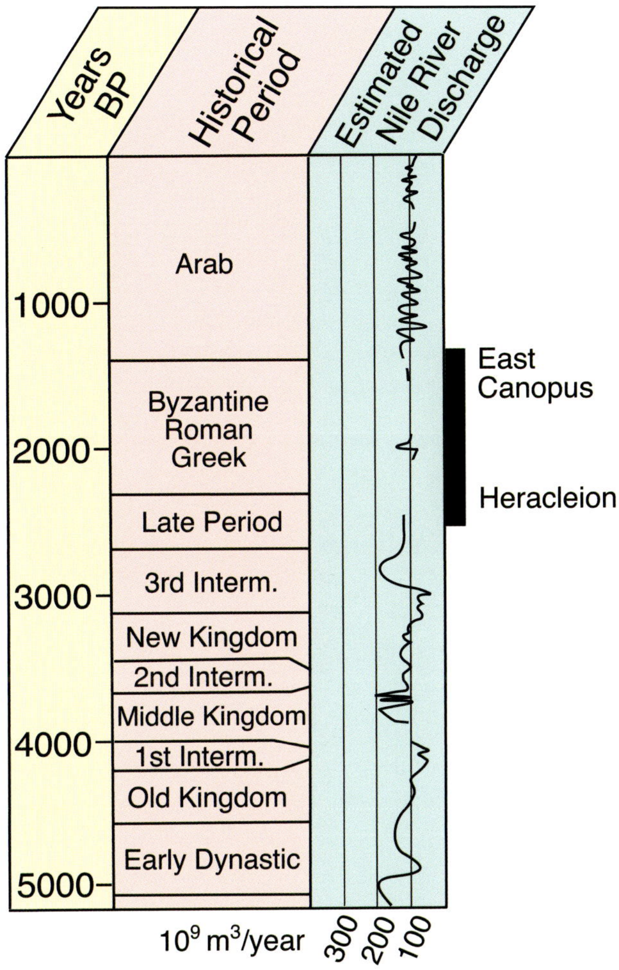

Figure 2.5 Summary of Nile river discharge variations during the middle to late Holocene. Largely derived from data in Said (1993).

Deposits of former Canopic distributaries and its now submerged promontory have been identified in radiocarbon-dated cores recovered on land, and also in bottom grab and core samples collected seaward of the present coast (Chen *et al.* 1992; Frihy 1992a,b; Frihy *et al.* 1994; Stanley *et al.*, Chapter 3 in this volume). Analyses of texture and sand composition in radiocarbon-dated sediment cores recovered near the coast indicate that the Canopic channel was active as early as 6000 years ago (Chen *et al.* 1992). At that time, when sea level was lower than at present, the Canopic promontory formed on what has now become west-central Aboukir Bay (Figure 2.10 A,B). Global and regional analyses

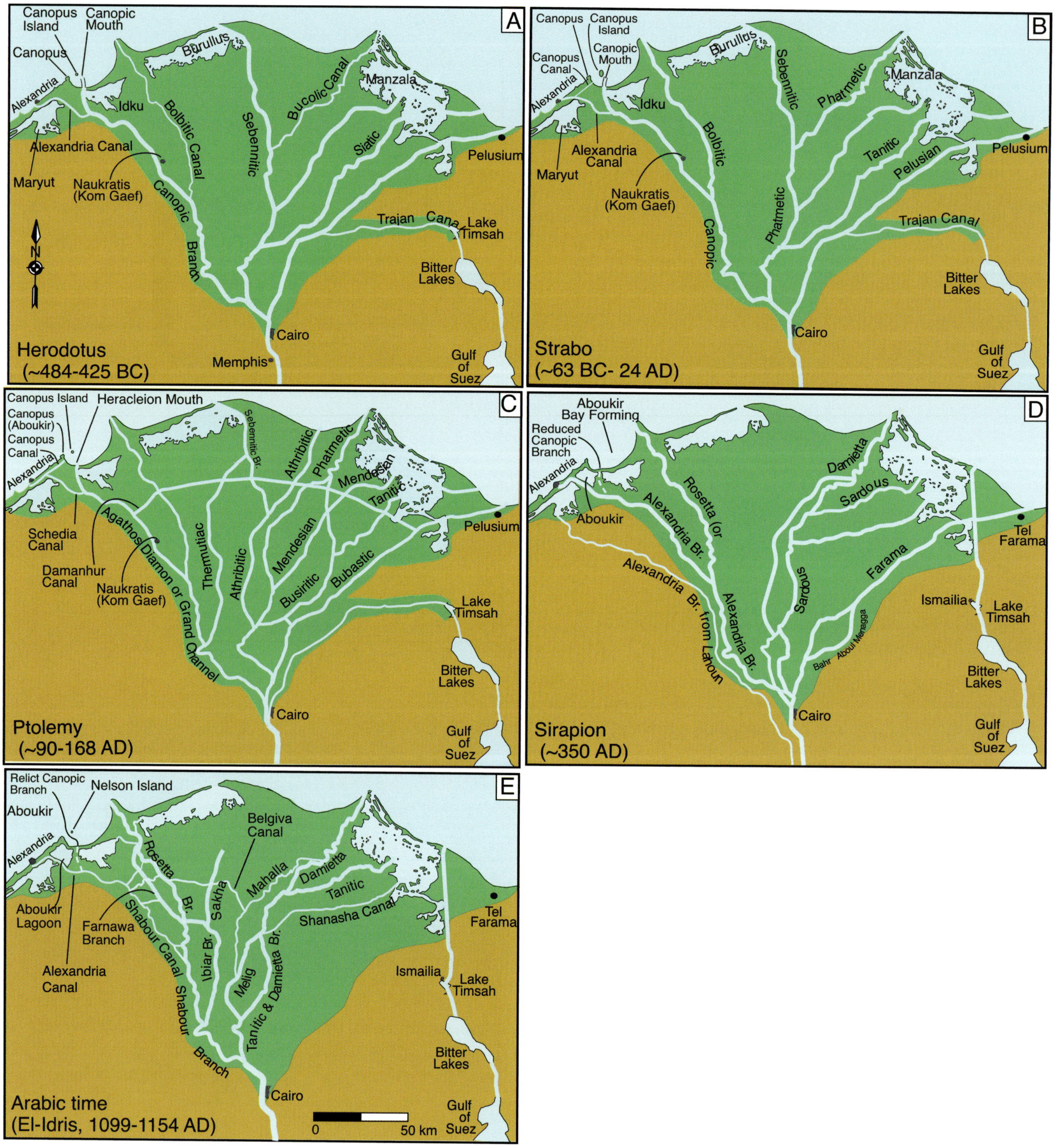

Figure 2.6 Generalised maps of the Nile delta showing the positions of major distributary channels and canals during the times of (A) Herodotos (~484–425 BC), (B) Strabo (~63 BC to AD 24), (C) Ptolemy (~AD 90–168), (D) Sirapion (~AD 350) and (E) Arabic era (~AD 1099–1154). Note that channel configurations are shown on an early twentieth-century base map of the delta, and hence these maps do not record change in configuration of the coastal region. Modified from Toussoun (1926).

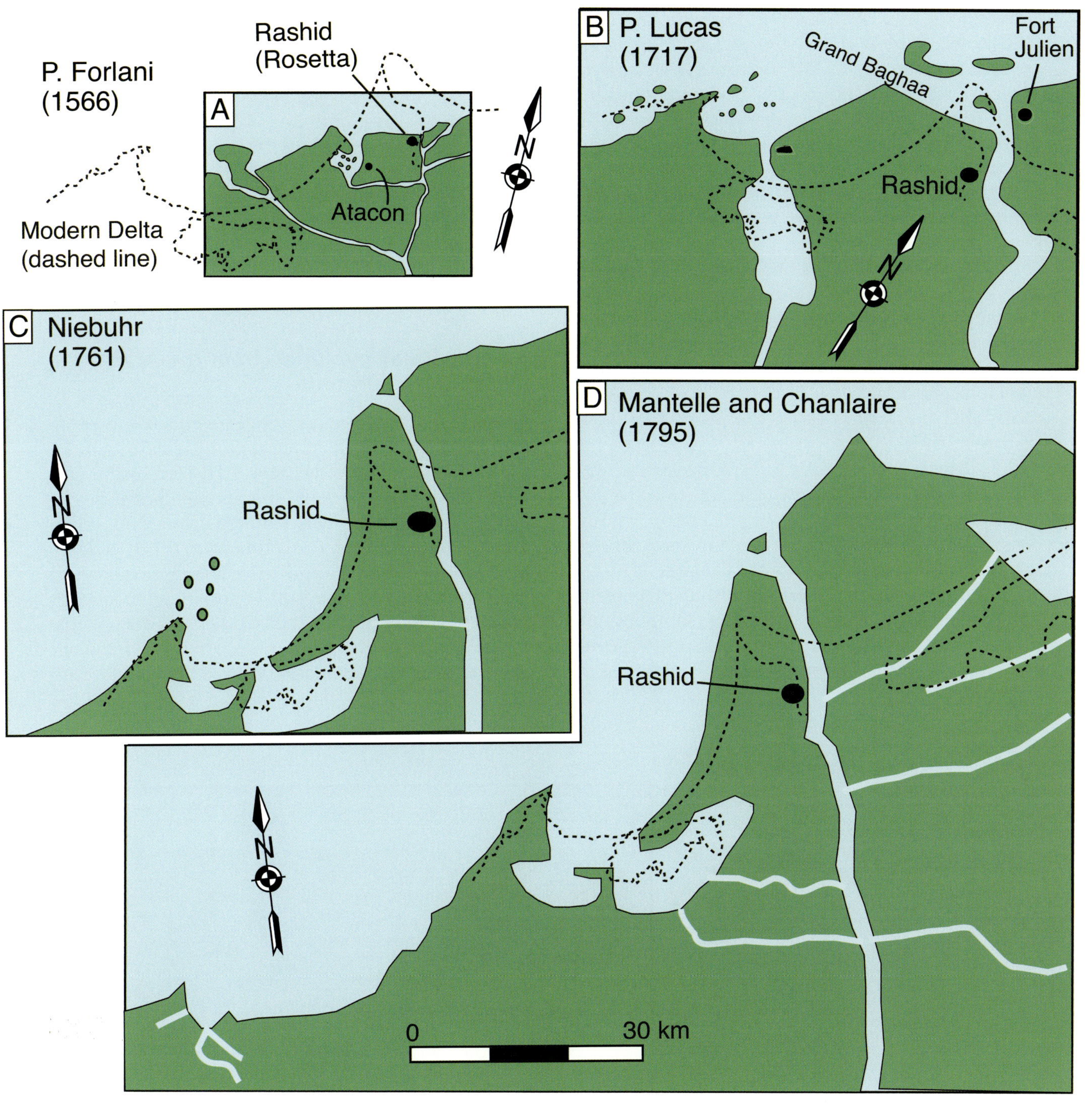

Figure 2.7 Historical maps showing the configuration of the north-west Nile delta from the middle sixteenth to the late eighteenth century AD. For B, C and D, the modern shoreline (dashed line) is superimposed for reference by matching the position of the Aboukir headland and the town of Rashid (Rosetta) on both historic and modern maps. Because the Aboukir peninsula was not included in historic map A, the scaling and orientation of this map (relative to the modern configuration) are more conjectural. The four maps are compiled from data in UNDP/UNESCO (1978).

indicate that sea level steadily rose at rates of ~1–3 mm per year for the period 600 BC to AD 800 (Fairbanks 1989; Pirazolli 1992).

Bathymetric surveys indicate that the Canopic promontory was more than 10 km wide along the delta coast, extended seaward at least 12 km, and covered an area of ~70 km^2 (Figure 2.3A). In terms of size, shape and surface area, the promontory of the former Canopic is comparable to that of the present Rosetta (Figure 2.3A). High-resolution seismic profiles collected in 2000 show that the Holocene delta sediment section thickens east of the Aboukir headland (Stanley *et al.*, in this volume). Core analysis, integrated with recent seismic surveys, indicates that unconsolidated, organic-rich silts and sands of the Canopic promontory were deposited on a consolidated carbonate substrate of late Pleistocene age. This limestone kurkar, offset by normal faults into a horst and graben topography in pre-Holocene time, was buried by carbonate and river Nile sediments during the past ~7000 years (Figure 2.10A).

Subsurface and recently acquired seismic data also indicate that, during the middle to late Holocene, the location of the Canopic mouth periodically shifted along the coast by as much as 5 km (Figure 2.10) (Stanley *et al.*, Chapter 3 in this volume). Short-term river mouth displacements typically involve crevasse-splay development, and characterise many modern delta systems (Coleman and Wright 1975; Coleman 1982; Penland *et al.* 1988). Similar shifts of the distributary channel have been documented in recent time at the Rosetta promontory (US Defense Mapping Agency 1977; Blodget *et al.* 1991) (Figure 2.11). As the Canopic mouth shifted, a series of small localised and temporal sub-promontories would have developed (Figure 2.10B,C). As the mouth continued to shift (over a period of 10–100 years), the sub-promontories would dissipate and subside beneath the steadily rising sea. Construction of large heavy stone structures, such as the temples of Heracleion and East Canopus, directly on the soft, water-saturated silt-rich sediments of the Canopic sub-promontories further promoted substrate instability and subsidence.

6 Aboukir Bay evolution from the Arabic period to the present

Integrated archaeological (Goddio F, 2007), geological and historical documentation indicate that, by the end of the eighth century AD, Heracleion and East Canopus were fully submerged in the developing Aboukir Bay. This resulted from land subsidence and sea level rise (Toussoun 1934; Stanley and Warne 1993a,b), and also was a response to local growth-fault and soft sediment deformation processes (Stanley *et al.* 2001, 2004). By early Arabic times, much of the Canopic channel system had been converted into an artificial canal and drain

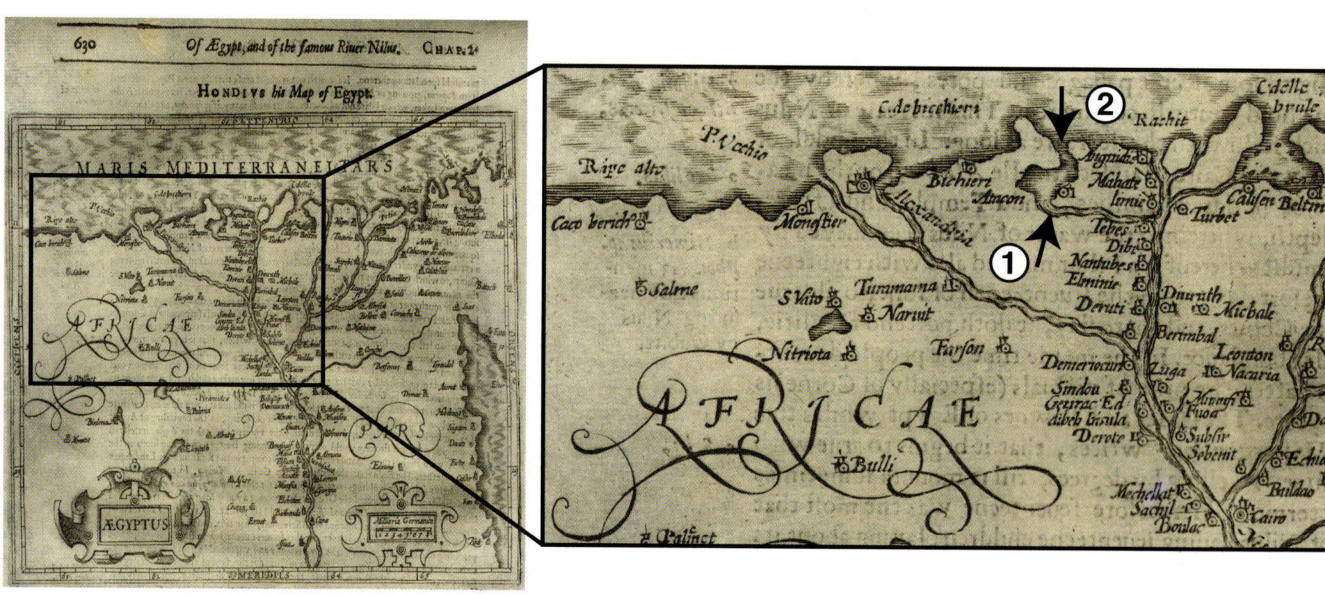

Figure 2.8 Map by Hondius (1625 BC) of the lower Nile (left), and an enlarged section of this map showing the north-western Nile delta (right). The relict Canopic branch (1) is shown extending to Edkou lagoon and the Mediterranean coast (2) west of Rosetta.

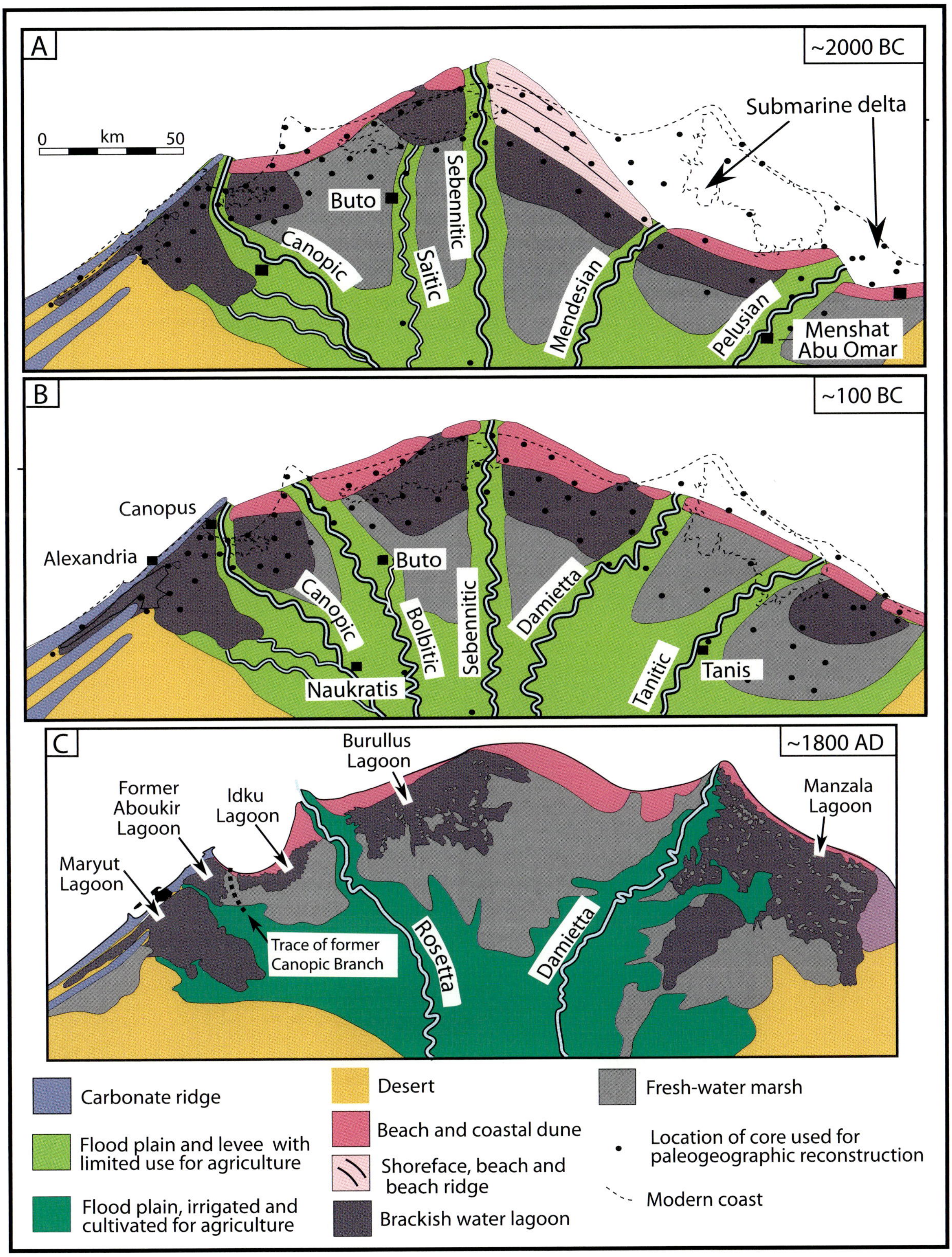

Figure 2.9 Palaeogeographical reconstruction of the northern Nile delta for (A) ~2000 BC, (B) ~100 BC, and (C) ~AD 1800. Interpretations for maps A and B are based on analysis of radiocarbon-dated sediment cores (modified from Stanley and Warne 1993a,b). Analysis emphasises that the shoreline position markedly changed as the different delta distributary channels waxed and waned. Map C is based on compilation during occupation of Egypt by Napoleon's army (modified from Jacotin 1826).

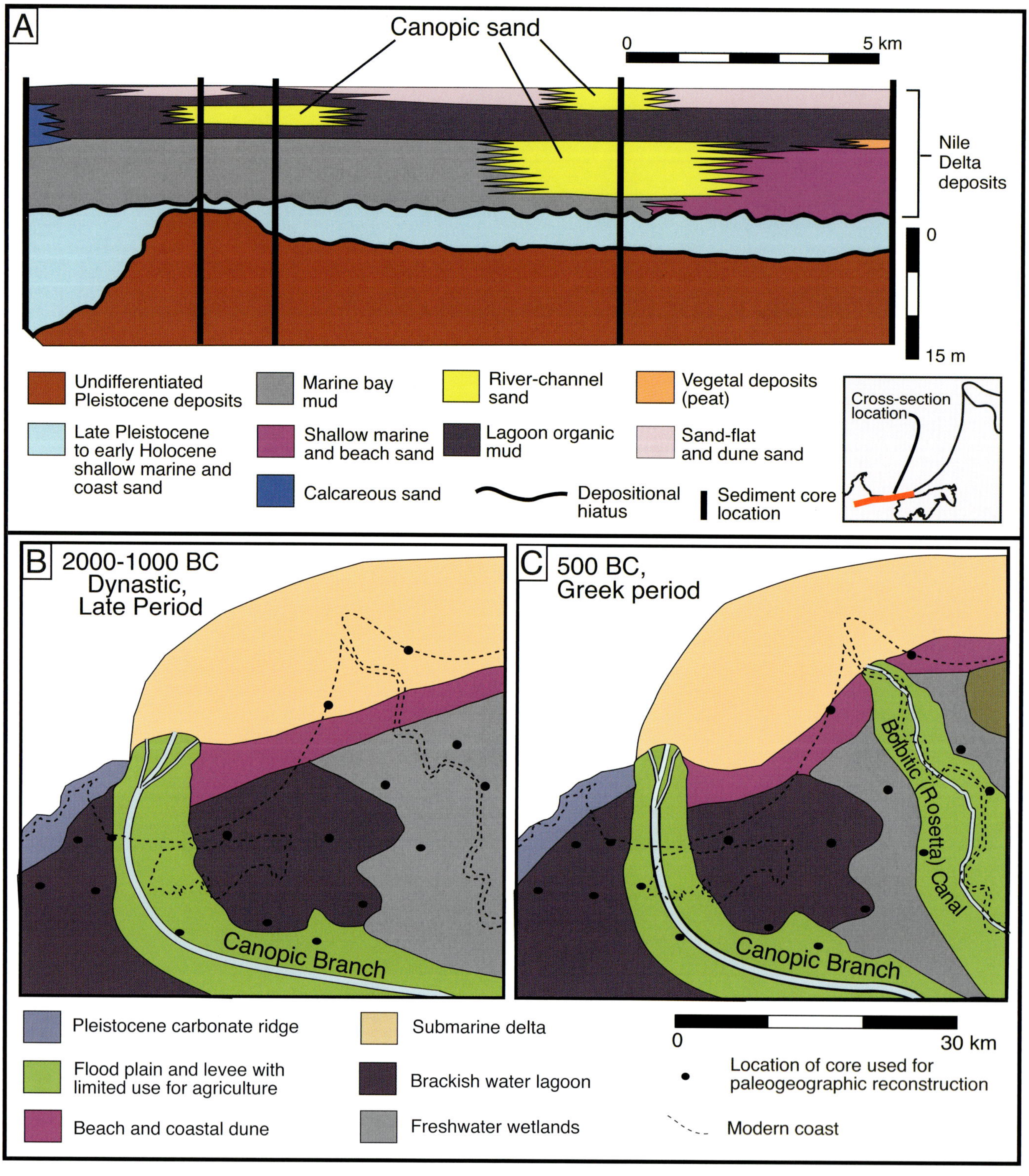

Figure 2.10 (A) Geological cross-section in the Aboukir region of the north-west Nile delta. This profile records changes in position of the Canopic branch over time. (B) Palaeogeographical reconstruction of the north-western delta at about 2000–1000 BC, based on stratigraphic analysis of radiocarbon-dated sediment cores. (C) Reconstruction at ~500 BC. Note that the Canopic channel and mouth migrated laterally over time; the Bolbitic (later Rosetta) branch had begun to develop by Greek times. Modified from Chen *et al.* (1992).

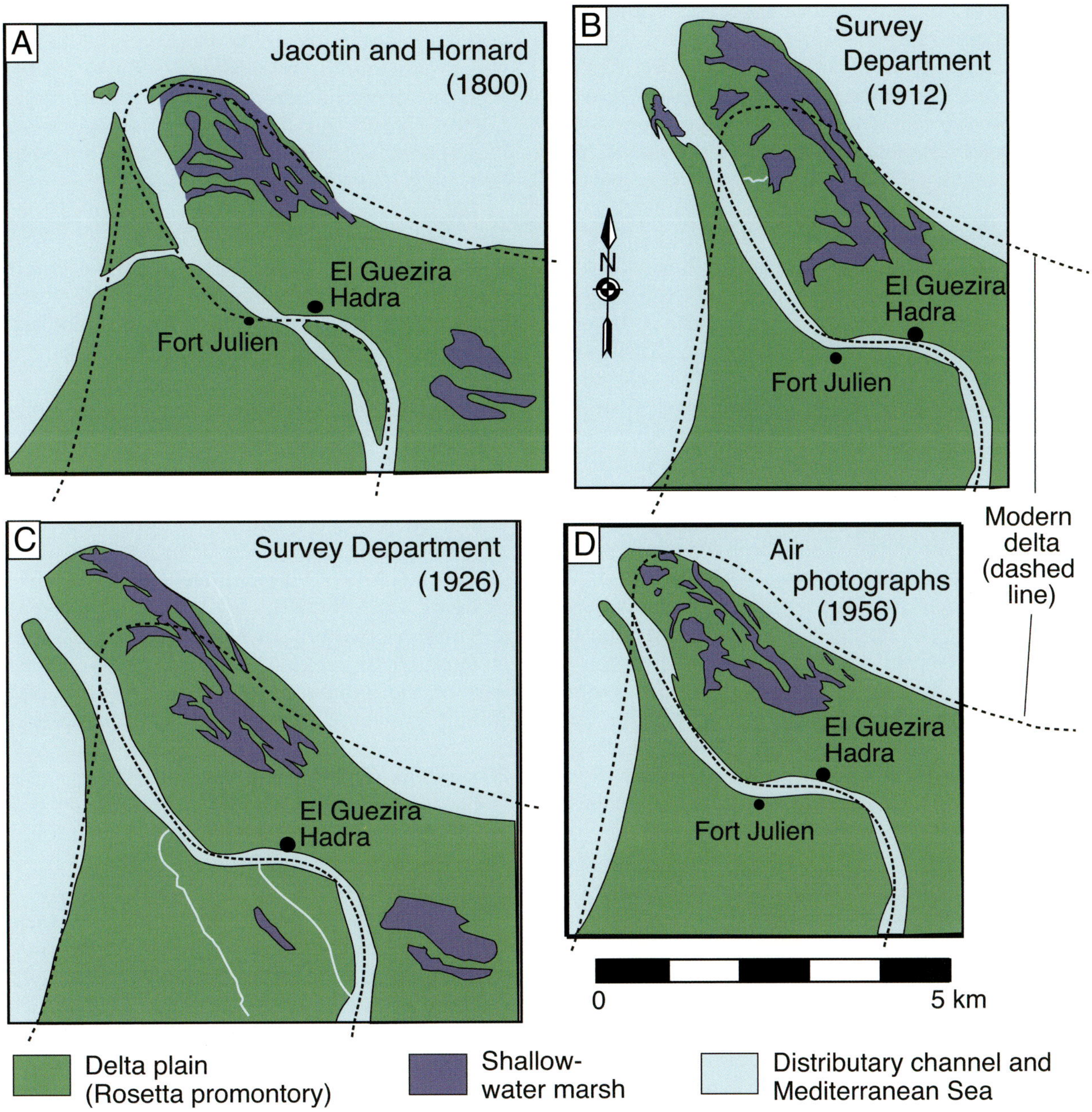

Figure 2.11 Historic maps showing the advance (A, B and C) and retreat (D) of the Rosetta promontory, north-western Nile delta (modified from UNDP/UNESCO 1978). These record rapid displacement of channel path and associated depressions, including small temporary wetland basins, and demonstrate the dynamic nature of this Nile distributary channel and mouth over a period of ~150 years. The retreat of the Rosetta promontory between 1926 and 1956 was caused largely by construction of the Aswan Low Dam and intensified irrigation and drainage projects in the delta plain.

system, while progressively more Nile water and sediment were discharged to the coast by the Bolbitic-Rosetta system (Figure 2.6E).

Before the end of the first millennium BC, the Nile distributary system was reduced to two major channels – the artificially maintained Rosetta and Damietta branches. The formation of Aboukir Bay resulted primarily from submergence of the Canopic promontory and rapid progradation of the Rosetta promontory. The Rosetta promontory caused partial refraction of east-directed littoral currents and formation of a clockwise eddy on the western side of the promontory (Figure 2.2). Littoral transport of sediment formed broad beach and dune deposits along the flanks of the Rosetta promontory. Bottom currents in the bay displaced both older reworked (relict Canopic) along with modern Rosetta-derived sediments on the shelf. Some of these sediments have accumulated preferentially in areas of weaker current activity, primarily towards the centre of Aboukir Bay, east of the submerged Canopic promontory, and also at greater depths farther to the north on the middle and outer Egyptian Shelf (Frihy *et al.* 1994) (Figure 2.3B).

As populations increased in lower Egypt, distributary channels were converted into irrigation canals and waste-water drains, some of which did not extend to the coast (De Cosson 1935; Shafei 1952; Stanley 1996). Development of a dense and complex irrigation network, combined with a decrease in annual Nile discharge, resulted in an overall decrease in flow through the Rosetta and Damietta distributaries. However, about 70% of Nile annual discharge to the coast has been through the Rosetta (Said 1981), resulting in the accumulation of large volumes of sediment at the Rosetta promontory. Aboukir and Edkou lagoons diminished in size (Figures 2.4, 2.10C) as the volume of fresh water carried by the Canopic channel decreased, and as man drained and infilled these wetlands for agriculture (Hoffman 1979; Wenke 1991).

Although the Bolbitic-Rosetta became the major channel discharging sediment to the sea, flow through the Canopic branch continued to the coast during the early part of the second millennium BC, as recorded by the Arab scholar El-Idrisi (Figure 2.6E). Numerous historic maps (Figures 2.7, 2.8) show that, until the mid eighteenth century AD, the Canopic – while much diminished – still channelled Nile water seaward (maps in Siliotti 1998, including those of Abraham Ortelius, 1570; Hondius, 1625; Benoît de Maillet, 1735; Richard Pococke, 1737; and Jean-Baptiste Bourguignon d'Anville, 1766). Relict traces of the Canopic were still identified as recently as the nineteenth century by Arrowsmith (1802, 1807), Du Bois-Aymé (1813), Jacotin (1818) and others (Figure 2.9C).

The texture and mineralogical composition of modern near-shore and bay-floor sand can still identify the position of the lower stretch of the Canopic channel (Figure 2.12) (El Bouseily and Frihy 1984; El Fattah and Frihy 1988; Frihy *et al.* 1994). Moreover, a trace of the Canopic branch has been detected in the delta south of Edkou lagoon by aerial photography (UNDP/UNESCO 1978) and satellite imagery (Abdel-Kader 1982; IWACO 1989). A series of point bars and kôms south of Edkou lagoon (Figure 2.4), interpreted as remnant point bars related to channel migration, also delineate the flow path history of this distributary.

The recent detailed bathymetric survey associated with ongoing archaeological work in Aboukir Bay (Goddio 2007, this series) clearly demarcates the submerged Canopic promontory. This feature is bounded to the west by the Aboukir headland, to the north by the carbonate extension of Aboukir ridge, and to the east by a well-defined linear depression and ridge system trending from the coast at El Maadiya, north-north-westward to the submerged carbonate ridge extension (Figures 2.12B, 2.13A). The depth of the Canopic promontory is remarkably uniform, with about 80% of the surface ranging from 5.5 to 6.5 m. Of note, however, are five shallow (<5 m) features (Figure 2.13B): three elongate (to ~2 km long), located at the eastern promontory margin, oriented north–south; and two rectangular to irregular depressions, in the central part of the submerged promontory. Also, several small (diameter <500 m) depressions, deeper than 7 m, are recorded, including one adjacent to Heracleion. The distinct low-relief, linear depressions and ridges along the promontory's eastern boundary form a gentle eastward slope from ~7 m at the promontory surface to a depth of ~11 m and deeper in the central bay.

No deeply incised (>1 m relief), well-defined, continuous channels are apparent on the submerged Canopic promontory surface in the western part of Aboukir Bay (Figure 2.13A). This is in contrast to several distinct linear, continuous, closely spaced, parallel, linear depressions (>1 m relief) preserved along the eastern promontory margin. These trend from Fort Hamra in the south to as far northward as the carbonate Aboukir ridge extension (Figures 2.12B, 2.13B); these depressions may be the product of faulting or slumping. The partial preservation of depressions in the central portion of the submerged promontory in this relatively high-energy, shallow shelf environment is surprising, because sea-floor currents in the western bay have partially eroded the bottom, have reworked sediments laterally, and have smoothed the Canopic promontory surface. We thus surmise that these depressions are remnant, partially filled,

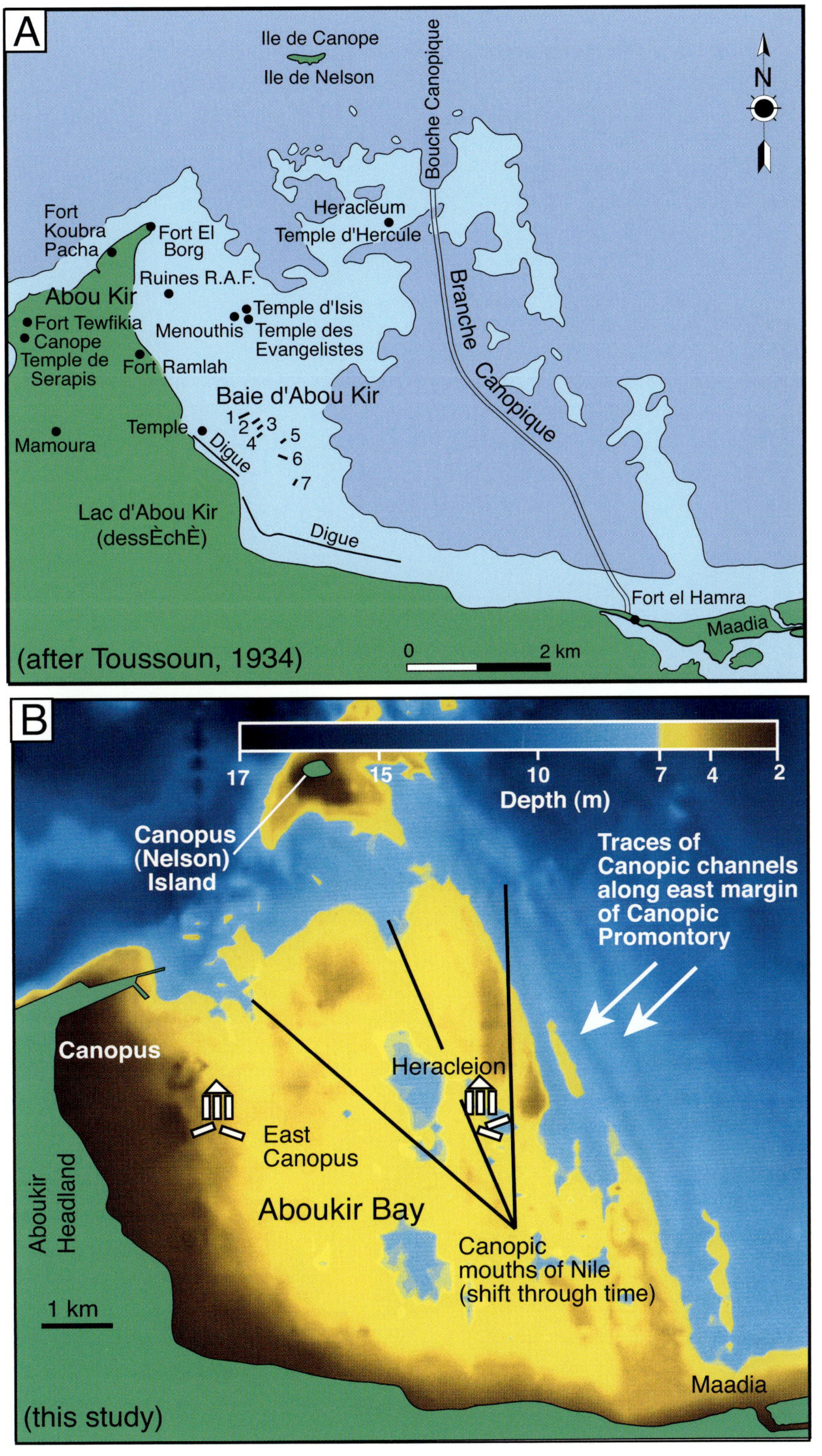

Figure 2.12 (A) Location of submerged archaeological sites and the Canopic branch in Aboukir Bay as interpreted by Toussoun in 1934. Note the submerged centres of Heracleion and Menouthis, and sites. (B) The centres of Heracleion (submerged during and/or shortly after the first century BC) and East Canopus (submerged at about 741–742 BC), located on the Canopic promontory, were mapped during the ongoing underwater exploratory study. Canopic channel courses, identified by tracing linear bathymetric lows on the present sea-floor, highlight the extent of migration of this former distributary through time. Topographic base provided by IEASM.

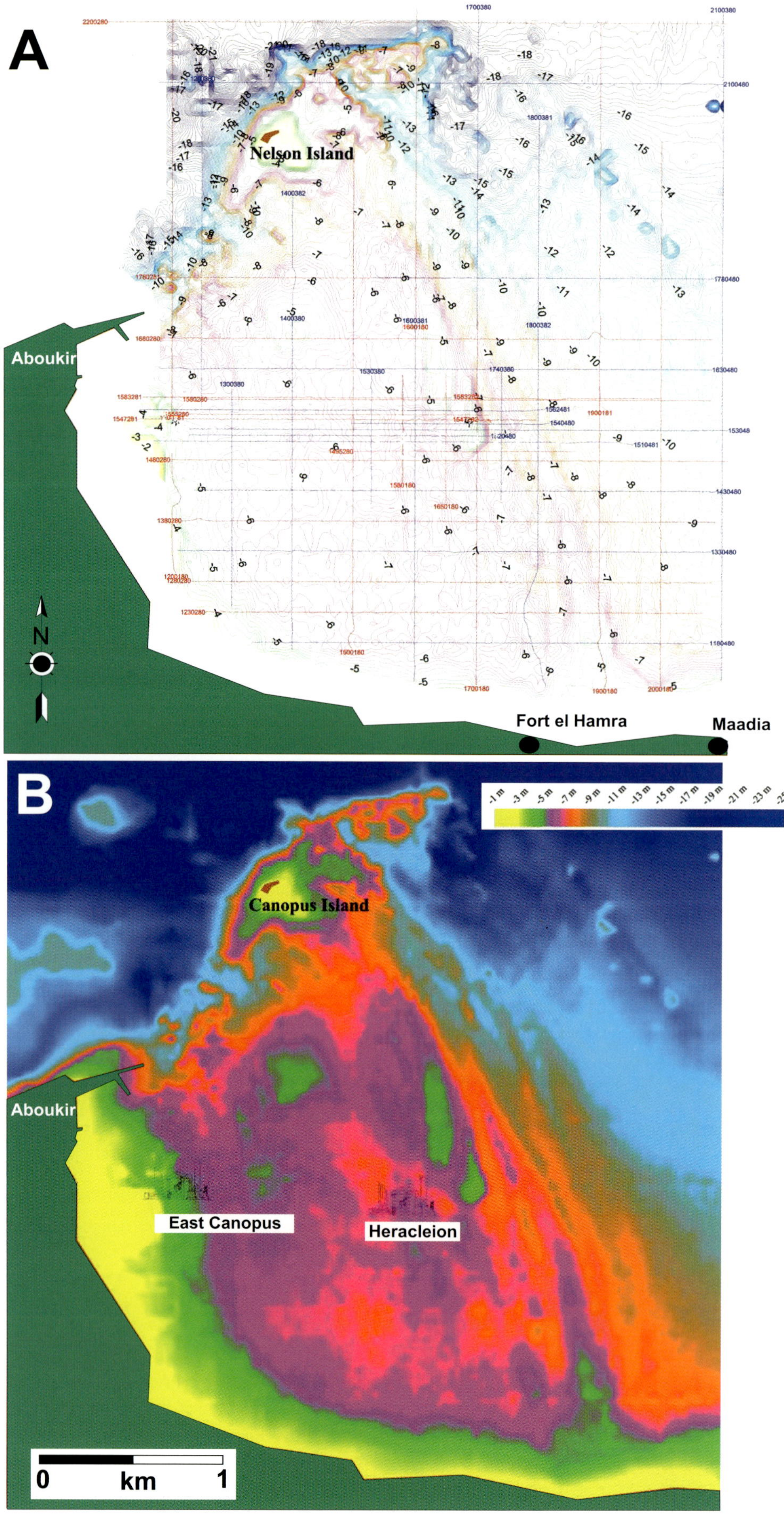
A
Nelson Island
Aboukir
N
Fort el Hamra
Maadia
B
-1 m
-3 m
-5 m
-7 m
-9 m
-11 m
-13 m
-15 m
-17 m
-19 m
-21 m
-23 m
-25 m
Canopus Island
Aboukir
East Canopus
Heracleion
0
km
1

Canopic channels, perhaps younger than those that once flowed on the shallower promontory.

Evidence of sediment truncation, reworking and depression fill is recorded by radiocarbon-dated grab and core samples of mid and late Holocene age recovered at and near the present surface of the promontory (Stanley *et al.* 2001, 2004; Chapter 3 in this volume). As visually observed by the mission excavating the sites in Aboukir Bay and from shallow cores (to ~5 m length), there is a thin layer (to 1 m thick) of recent shelly marine sand that partially covers the much older, dark organic, silt-rich sediment that forms the submerged promontory substrate.

7 Summary and conclusions

Current sea-floor exploration and archaeological excavation show that the two sites, each covering a surface area of nearly a square kilometre, presently lie at water depths ranging from about 5 to 7 m. This indicates submergence of the structures during and following their occupation. Global sea level rise since the eighth century BC has been less than 2 m along the Nile delta coast. Therefore, substrate compaction, tectonic subsidence, erosion and perhaps other phenomena must account for the remaining 3–5 m of submergence since Byzantine times. More extensive information on this topic is provided in Chapter 3 of this volume.

On the basis of analysis and integration of historic records, remote sensing imagery and our recent geological data, we summarise findings as follows:

- Coastal changes were largely a function of the relative rate of river sediment input and coastal erosion (Figure 2.10B,C); the form and elevation of the delta coast were also modified by dynamic interaction of sea level rise, land subsidence and sediment reworking by waves and wind.
- Heracleion was positioned near the mouth of the Canopic branch on the low-elevation, but subaerially exposed, Canopic promontory (Figures 2.9A,B, 2.10B,C).
- The shoreline during occupation of the two sites was located north of the present coast, in what is now Aboukir Bay.
- Extensive lagoons and marshes were present landward of the Nile delta coast and the Canopic promontory (Figures 2.9A,B, 2.10B,C).
- The delta plain environment of the Canopic promontory was dynamic (Figures 2.7–2.10): decadal- to centennial-scale climatic cycles, together with variations in annual floods, caused frequent, sometimes sudden, shifts in the position of the Canopic channel mouth.
- Shifts in channel mouth position caused development of localised, laterally migrating sub-promontories; when the channel mouth switched and abandoned a sub-promontory, the latter began to submerge, was reworked into beach sands, or was covered by aeolian dune deposits.
- Periodic shifts of the Canopic channel, its mouth and the associated sub-promontory caused substantial changes in the depositional environments around Heracleion and East Canopus, including the position of these sites relative to the main channel and to the coast.
- It is likely that Heracleion was originally positioned on coastal sand bars and/or dunes that were little more than 1–2 m above sea level (Figure 2.1); however, coring shows that the substrate beneath these relatively thin silt and

Facing page: **Figure 2.13** (A) Map of the sea-floor in present Aboukir Bay and location of Heracleion and East Canopus (Courtesy of IEASM). The map is based on north–south and east–west grid lines, with spacings of 100–1000 m. Contours (interval = 20 cm) detail the submerged, low-relief Canopic promontory surface and north–south, east–west and south-east to north-west trends on the promontory surface. The eastern boundary of the promontory is marked by distinct linear south-south-east to north-north-west channels extending from the town of El Maadiya on the coast northward to the Aboukir carbonate extension. (B) Map, simplified from (A), depicting less distinct depressions, probably partially filled channels that once extended from Fort Hamra at the coast (the historically documented site of the Canopic branch), across the promontory, towards the sites of Heracleion and East Canopus. Also apparent on the promontory are well-defined highs (<5 m) and several small depressions (>7m), including one adjacent to Heracleion. Maps A and B were generated by bathymetric surveys conducted in April 2001, as part of the IEASM underwater archaeological survey of Heracleion and Canopus. The bathymetric survey was conducted using a high-resolution single-beam echosounder TRITECH PA500 (IEASM). The north–south and east–west grid spacings (ranging from 100 to 1000 m) used for the survey were determined using real-time differential GPS.

sand units comprised mostly soft, water-saturated mud (Figure 2.10A).

- Emplacement of fixed, massive structures, including stone temples, on the metastable, organic-rich mud forming the sub-promontories, promoted instability and rapid subsidence.

- After the establishment of Heracleion and East Canopus, natural and weight-induced subsidence would have rendered these sites progressively more vulnerable to the annual floods of the river Nile, especially those floods reaching a stage of 1 m or more above average high flood level.

- In addition to Nile floods, earthquake tremors and tsunamis may have also influenced the integrity of Heracleion and East Canopus, lying just above sea level on an unstable, soft sediment substrate; no records, however, document the impact of seismic events on these two sites.

- Since the submergence of Heracleion and East Canopus at the end of the eighth century AD, the configuration of the north-west Nile delta has fundamentally changed, largely because of decrease in discharge through the Canopic branch and marked increase in discharge to the east through the Bolbitic-Rosetta branch. The altered proportion of Nile flow and sediment released through these adjacent distributary channels, coupled with land subsidence and sea level rise, resulted in submergence of the Canopic promontory, seaward development of the Rosetta promontory and, ultimately, development of Aboukir Bay (Figures 2.3, 2.9, 2.10, 2.13).

Chapter 3

Submergence of Archaeological Sites in Aboukir Bay, the Result of Gradual Long-term Processes plus Catastrophic Events

Jean-Daniel Stanley, Gérard Schnepp and Thomas Jorstad

'When the Nile comes out upon the country, only the cities show above the watery surface, very much like islands in the Aegean Sea.' Herodotus, *The Histories*, 2.97

1 Introduction

Heracleion and East Canopus were important Egyptian settlements located on the Mediterranean coast of Egypt. They lay along Egypt's north-western Nile delta, but their recent rediscovery by underwater archaeological exploration shows that they are now submerged at depths of 5–7 m in Aboukir Bay. As archaeologically (Goddio F, 2007) and geologically documented in this volume, the two cities were vital gateways to Egypt, serving as transit ports for navigation to the Greek trading centre of Naukratis positioned in the delta proper about 55 km inland from the present coast. The Egyptian cities subsequently became Ptolemaic, then Roman and finally Byzantine centres, and they disappeared by the time of, or shortly after, the Arab conquest in the mid-seventh century AD. Scholars have done little more than speculate that the settlements may have been logistically related to each other. They generally agree, however, that the two cities were once located at, or close to, the mouths of the now defunct Canopic distributary branches of the Nile (Bernand 1970).

The navigational activity at Heracleion, also known as Thonis to the Egyptians (F. Goddio in Constanty 2002; Goddio 2007; also in IEASM maps and documents) and East Canopus thrived when the Canopic branch was still the largest of the Nile distributaries flowing in the north-western Nile delta (Toussoun 1922; Stanley and Warne 1993a, 1998), and when the Aboukir Bay shoreline was about 5 km north of the present coast (Stanley and Warne, Chapter 2 in this volume). Ongoing exploration indicates that these two centres were active over a long period, from the eighth century BC until the eighth century AD.

Questions raised by historians and geographers as to where and when these cities disappeared into the western part of Aboukir Bay have remained, until recently, without satisfactory answers. Even less is known about the processes that caused the subsidence of these two cities, and quite possibly other settlements yet to be discovered beneath the waters of the bay. In this chapter, we present newly acquired physical evidence indicating that the demise of both Heracleion and East Canopus was primarily the result of their construction on soft, unstable, water-saturated sediment located along the annually flooded and periodically shifting river channels of Canopic deltaic sub-lobes that developed east of Canopus. Herein, special attention is paid to the identification of the natural processes that would have triggered site failure, and to determine whether these phenomena were gradual occurring over a long period, or the result of sudden and powerful events of short duration, or a combination of the two.

2 The problem of site submergence

Of the various scholars concerned with the disappearance of historic sites in western Aboukir Bay (Figures 3.1A, 3.6), the Egyptian Prince Omar Toussoun most directly focused on the possible causes of their submergence. Toussoun's interpretations were based on a compilation of historical and geographical documents, on aerial flight reconnaissance of shallow near-shore areas made in the early 1930s by the Royal Air Force, and on a most productive 'hard hat' dive that he organised in the bay on 5 May 1933. Shortly afterwards, he presented the results of his research at the Société Royale d'Archéologie (Toussoun 1934). His article included a map showing the location of various submerged archaeological sites and the geographical position of the former Canopic channel relative to the present Aboukir peninsula (the map is reproduced as Figure 2.12 in this volume). Also of note in his discussion were several possible mechanisms that could have led to site submergence. For this analysis, Toussoun took into account the writings of Sophronios of Jerusalem, dating from the beginning of the seventh century BC. His writings indicate that, before its demise, the temple of the Evangelists was located in East Canopus. Sophronios describes some remains of the temple located on a beach, covered with sand, with marine water lapping against them. Thus, the temple's foundations final submergence are dated to a time after this early seventh century AD report by Sophronios, i.e. either in late Byzantine time or in the period of early Arab conquest in Egypt.

It is our view that Toussoun and most other historians (cited in Bernand 1970) who evaluated the demise of the sites have correctly interpreted submergence of the settlements in the bay as having resulted primarily from natural rather than from anthropogenic causes. Scholars, however, have remained truly perplexed by the lack of any writings describing the final sequence of events that led to the disappearance of Heracleion, East Canopus, and the channel configuration of the Canopic branch east of the Aboukir peninsula. Until now, little attention has been paid to the identification of specific natural mechanisms responsible for the disappearance of the cities, and whether these were gradual, long-term natural events, or of a sudden and catastrophic nature (Nur 2000; Stanley *et al.* 2001, 2004; Constanty 2002). Our goal here, then, is to describe the pertinent physical evidence recently collected to formulate rational working hypotheses pertaining to how and why site abandonment and subsidence occurred.

The combined archaeological and geological exploratory efforts initiated by Franck Goddio and the European Institute for Submarine Archaeology (IEASM) team between the late 1990s and the present make it possible to advance beyond the initial inferences of Toussoun (1934). Toussoun suggested that the lowering of the land and the rise of sea level may have been major factors leading to site submergence in this region since the sixth century AD. One question to be asked at this time is the following: To what extent would gradual natural causes and sudden catastrophic events have significantly influenced the evolution and development of the bay and have caused foundering of the cities during the past 2500 years? For this, it is essential to distinguish the evidence of processes active over the long term from more sudden, intense and destructive events such as powerful Nile floods, earthquakes, tsunamis, storm wave surges and episodic sudden failure of unstable sediment sections (such as by massive slumping) on the bay-floor in areas on and proximal to the Canopic delta sub-lobes.

Since available historic documents do not shed light on submergence of the two centres, the ongoing fieldwork by the IEASM team has undertaken geoarchaeological studies that closely couple the physical and archaeological sciences. The causes of destruction, subsidence and disappearance of the historic settlements in Aboukir Bay take into account archaeological information, but are thus evaluated here primarily in light of geological, sedimentological, hydrological, soil engineering and palaeoenvironmental considerations.

3 Geological setting of the study area

During the past several millennia, the north-western Nile delta margin has evolved on a geological terrain termed the North Delta Block. This is a tectonically lowered sector that lies above and northward of an underlying zone of major normal and growth fault structures. The vertically offset area, separated from the Nile delta terrain to the south (South Delta Block) by an east–west oriented structural Hinge Line (Figure 3.1B), resulted from the subsidence of a thick depositional sequence of consolidated Tertiary units (in Figure 3.1C, see the zone underlying petroleum wells nos. 4–7). This lowered North Block sequence comprises nearly 5000 m of partially to well-consolidated strata ranging from the Mesozoic to the Quaternary age (Figure 3.1C). Deep boreholes drilled for oil and gas exploration in the North Delta Block have recovered carbonates, terrigenous deposits and anhydrites (Said 1981, 1990, 1993; Deibis 1982; Schlumberger 1984; Hussein and Abd-Allah 2001). The underlying faulted zone of Mesozoic to Miocene units is covered by somewhat less tectonically offset deposits of Pliocene and Quaternary age, ranging to ~2500 m in thickness. The thin (<5 m) surficial sediment veneer of Holocene age on the present western bay-floor is mostly recent Canopic deltaic sub-lobe deposits that have been eroded and reworked into a generally smooth, current-swept surface (Stanley and Warne Chapter 2 in this volume).

Geophysical surveys for petroleum exploration in this region have obtained deep seismic profiles (to 5 seconds acoustic penetration, equivalent to a thickness of nearly 5000 m) that indicate the western part of Aboukir Bay lies partially within a former tectonically active and faulted pull-apart basin (Figures 3.1C, 3.2A). The major structural trends in the north-west Nile delta and contiguous offshore region are oriented, for the most part, in a north-east to south-west direction (Figure 3.1B). This is in contrast to the area of the north-central delta, where major faults are oriented primarily along north–south trends. Quaternary to recent sediment of the upper 500 m of section in the western bay is generally continuous; these layers have been subject to offset only locally and, for the most part, to a minor extent (Figure 3.2B, arrow). Some shallow fault structures (growth, rotational types) are recorded along the delta margin and seaward off the north-west Nile delta. These affected primarily the upper several hundreds of metres or less of the unconsolidated sediment section (Ross and Uchupi 1977). These latter shallow, syndepositional structural systems are probably closely linked with the rapid accumulation, instability and failure of near-shore sediment sequences that once formed the Quaternary depositional units in this region. Such recent faulting, contemporaneous with rates of rapid deposition, is of the type commonly observed on large deltas on the world's coastal margins (examples in Coleman 1982, 1988; Maestro *et al.* 2002).

The formation of what is now the submerged Aboukir Bay floor is attributed to earlier faulting and pull-apart basin development in this region (Figures 3.1B,C, 3.2A) and to recent ongoing isostatic loading and sediment compaction. These effects result in part from the large depositional input at the mouth of the Canopic and, more recently, Rosetta branches of the Nile River. Analyses of late Quaternary sediment borehole data (Chen *et al.* 1992; Frihy 1992a,b; Warne and Stanley 1993; Stanley and Warne 1998) and geophysical

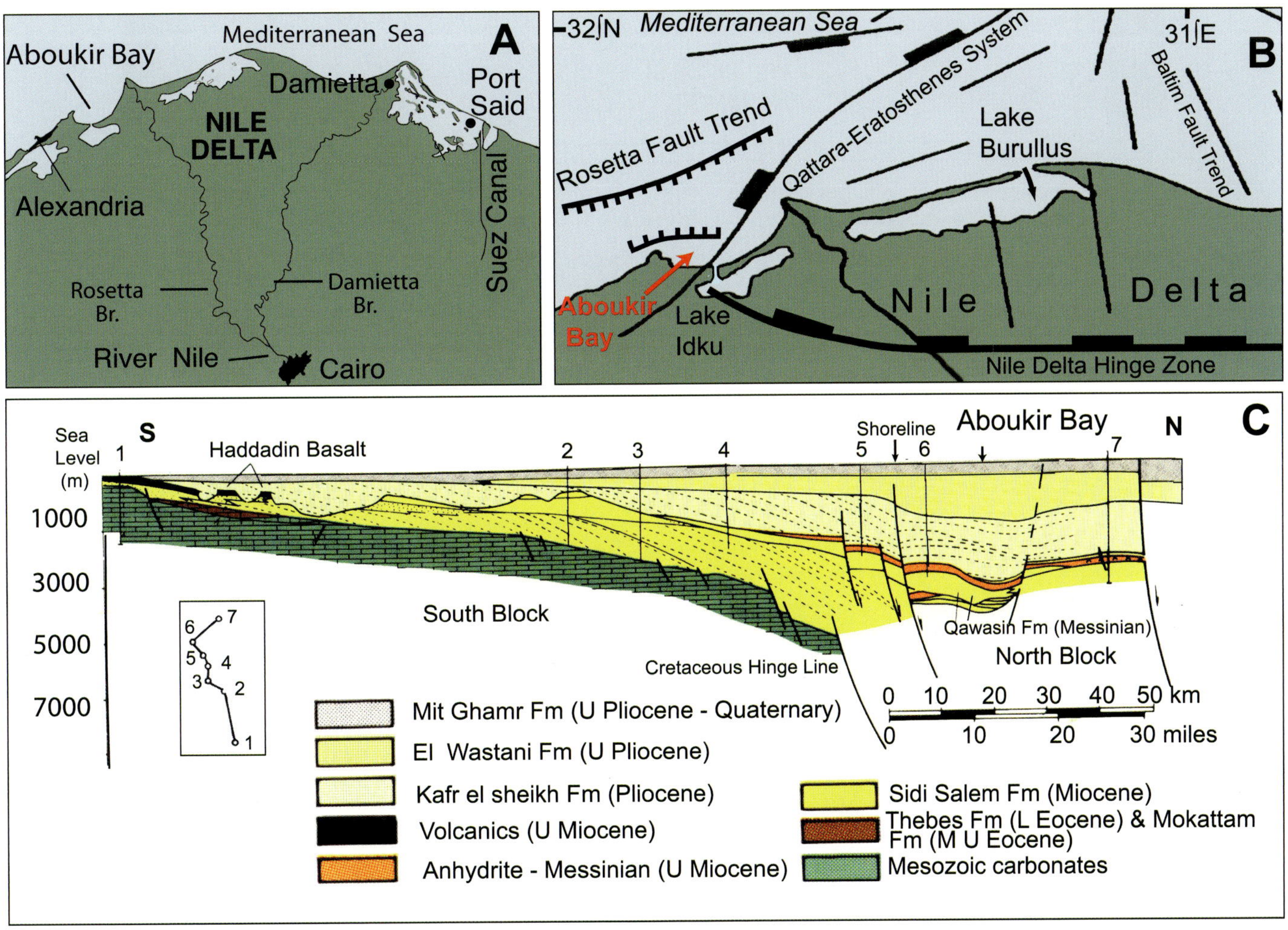

Figure 3.1 (A) Map showing Aboukir Bay on the north-western margin of the Nile delta, Egypt. (B) Major geological structures on the north-western and north-central margins of the Nile delta (modified after Said 1990; Hussein and Abd-Allah 2001, and other sources). (C) South–north transect across the north-western Nile delta margin, highlighting the major stratigraphic and structural configuration, including the graben-like basin underlying Quaternary units in Aboukir Bay (from Schlumberger 1984).

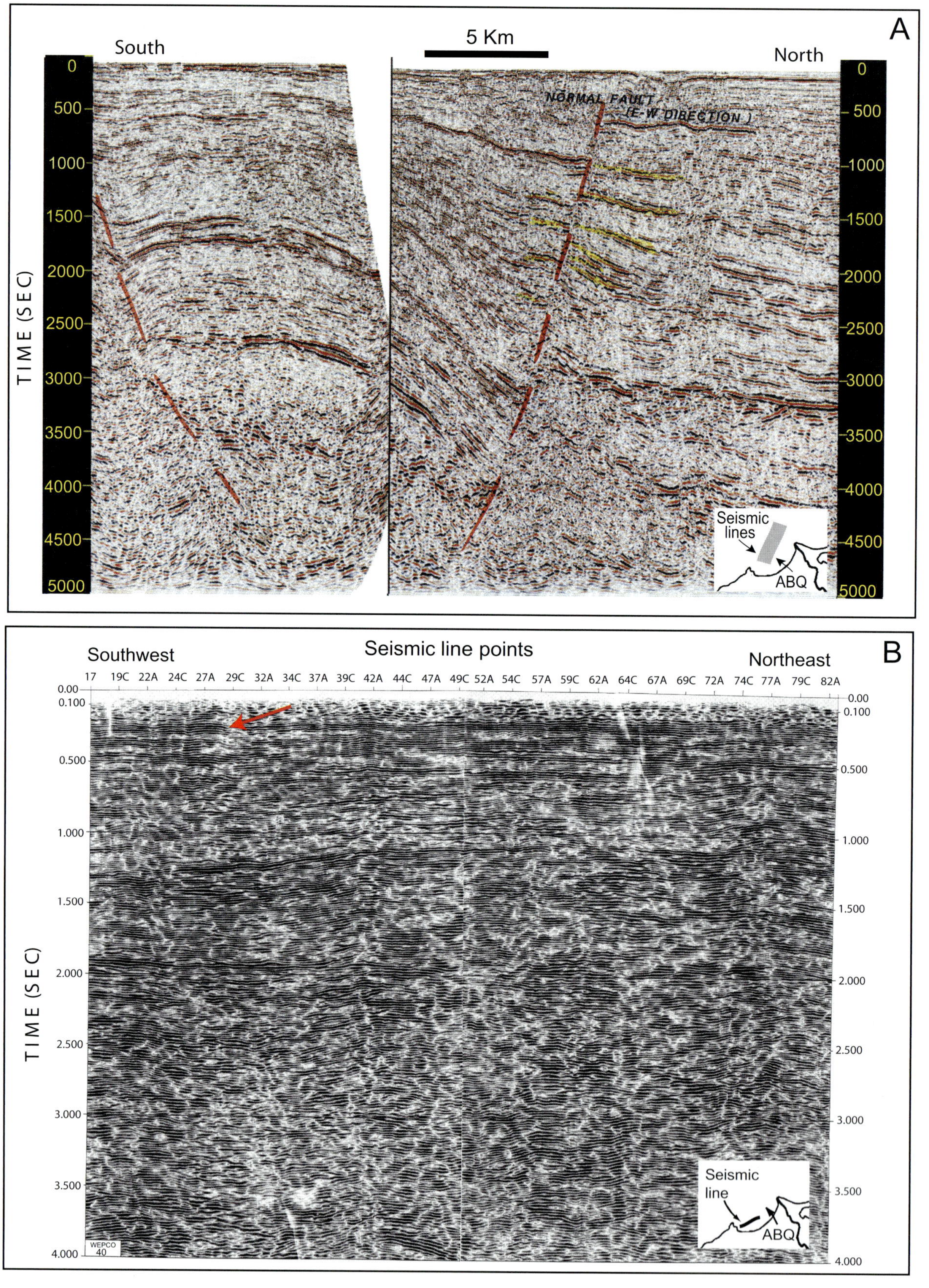

Figure 3.2 Seismic profiles in Aboukir Bay. (A) SSW–NNE transect, showing graben-like structural feature underlying Quaternary units in the north-central bay area. (B) South-west to north-east transect, in a sector where late Quaternary units appear generally continuous. A small offset in the western bay is shown by the red arrow. Depth given in seconds, two-way travel time.

analyses (Ross and Uchupi 1977; petroleum exploration records and reports) suggest that the bay-floor study area lies in a zone presently subject to relatively minor subsidence. This is substantiated by recent earthquake epicentre data (Kebeasy 1990), indicating that Aboukir Bay does not occupy a particularly highly active tectonic area (Figure 3.3A–C). The seismicity record for the bay proper (primarily micro-earthquakes: Kebeasy 1990; Stanley 1997) is in marked contrast to that of the Gulf of Suez and Levant region to the east and, especially, of the north-east Mediterranean (Figure 3.3D,E). Historically, earthquake activity has been recorded in the region to the west of the bay, in the vicinity of Alexandria (Figure 3.3A–C), that is located in the critical transitional zone between the Nile delta and the carbonate non-delta area to the west (UNDP/UNESCO 1978; Hassouba 1980).

The mean annual rate of vertical land displacement in this region has been calculated for the period from the early Holocene to the present. Land subsidence rates are based largely on radiocarbon-dated sediment core data (Stanley and Warne 1993a,b; Stanley *et al.* 1996), and these calculations take into account both world (eustatic) rise of sea level and regional lowering of land. Only modest subsidence values for the Holocene are recorded (Chen *et al.* 1992; Warne and Stanley 1993), i.e. average annual rates to as little as 0.1 mm/yr along some sectors of the coast of western Aboukir Bay (Figure 3.4). This finding is in marked contrast to the high measurements in the north-east Nile delta where rates of Holocene subsidence (to ~5.0 mm per year) are much higher (Stanley 1988; Stanley and Warne 1998). Whether lowering of the sites occurred in a gradual manner, or was caused by a series of

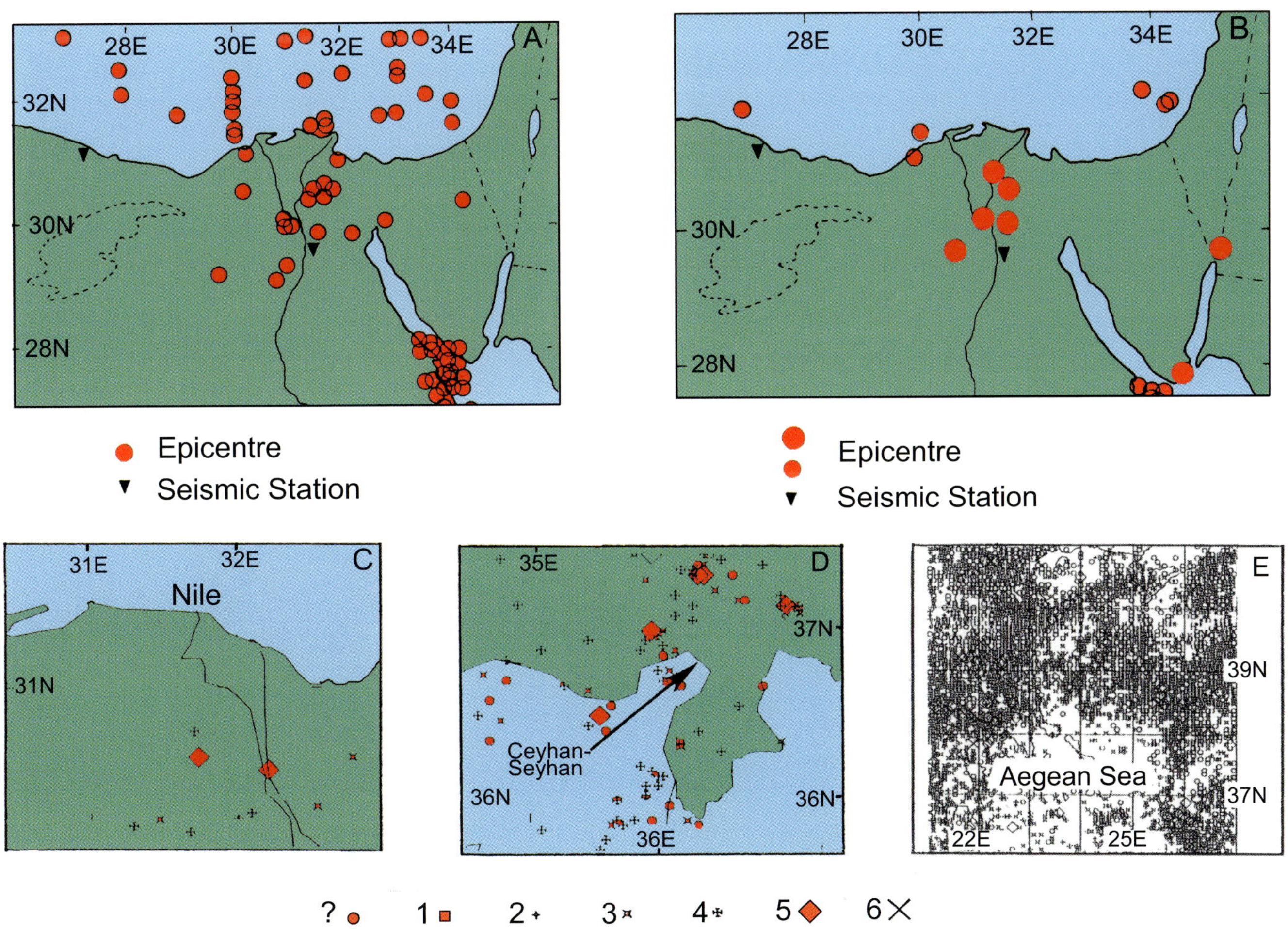

Figure 3.3 Earthquake epicentres in the eastern Mediterranean (A and B, from Kebeasy 1990; C–E, from Stanley 1997). (A) Location of seismic stations and epicentres of historical and recent medium to large earthquakes. (B) Epicentres of small earthquakes. (C) Epicentres in the Nile delta region recorded from April 1974 to October 1996. (D) Epicentres in the Ceyhan-Seyhan delta area of Turkey (arrow pointing to Gulf of Iskenderun) recorded from March 1945 to June 1996. (E) Epicentres in the Aegean Sea recorded from March 1933 to December 1996. Legend of epicentre magnitude (Richter scale) for panels C–E shown at bottom.

more irregular, sudden and large-scale catastrophic pulses, or both, will be explored later in this chapter.

4 Recent displacement of surficial strata

The initiation of a field and laboratory research programme to detail the configuration and petrology of the unconsolidated sediment substrate below the man-made structures was essential in order to understand how the two cities were lowered in western Aboukir Bay. Bathymetric, gravimetric and side-scan sonar exploration (conducted mostly in 1996 and 1997 by the IEASM), coupled with a high-resolution sub-bottom seismic survey (in April and May 2000) and a sediment vibracore recovery programme (in April and May 2001), were undertaken to map attributes of the Holocene deposits in the study area.

Cores previously collected in the bay proper (Frihy 1992a,b) and along the bay coast (Stanley *et al.* 1996) have shown that the Holocene sediment section ranges in thickness from ~3 m to possibly as much as 15 m. These geologically recent deposits, dated to the past ~7500 years (Stanley and Warne 1994), lie above much older units of Pleistocene age (>10,000 yrs). The Holocene veneer is comprised of mostly Nile-derived terrigenous (quartz-rich) material, to which has been added some carbonate particles transported to the bay from regions west of the Aboukir peninsula (Butzer 1960; El-Wakeel and El Sayed 1978; Alexandersson 1990; Chen *et al.* 1992; Frihy 1992a,b; Stanley and Hamza 1992; Warne and Stanley 1993). Cores indicate that the uppermost bay sediment section is a thin (<1 m to absent) surficial cover of modern marine sand formed largely of quartz and feldspar of medium sand size (>2.5 mm mean grain diameter) of Nile origin, and variable proportions of shell components such as mollusc and other biogenic debris. Some fine sand, silt and clay are derived from aeolian transport, but represent only a minor component of the surficial sediment.

Particularly valuable for this investigation of the sediment cover are seismic sub-bottom profiles obtained using a Triton EdgeTech XStar system in April and May 2000. The profiles, collected along 15 north–south and 22 east–west oriented transects, are spaced from about 100 to 1000 m apart. These account for a total of about 350 km of profiles covering an area of nearly 100 km^2 (Figure 3.5). Eighteen of the east–west profiles, some to a length of ~9 km and penetrating to a depth of ~10–15 ms (two-way travel time), or ~7.5–10 m below the water–sediment interface on the bay-floor, are

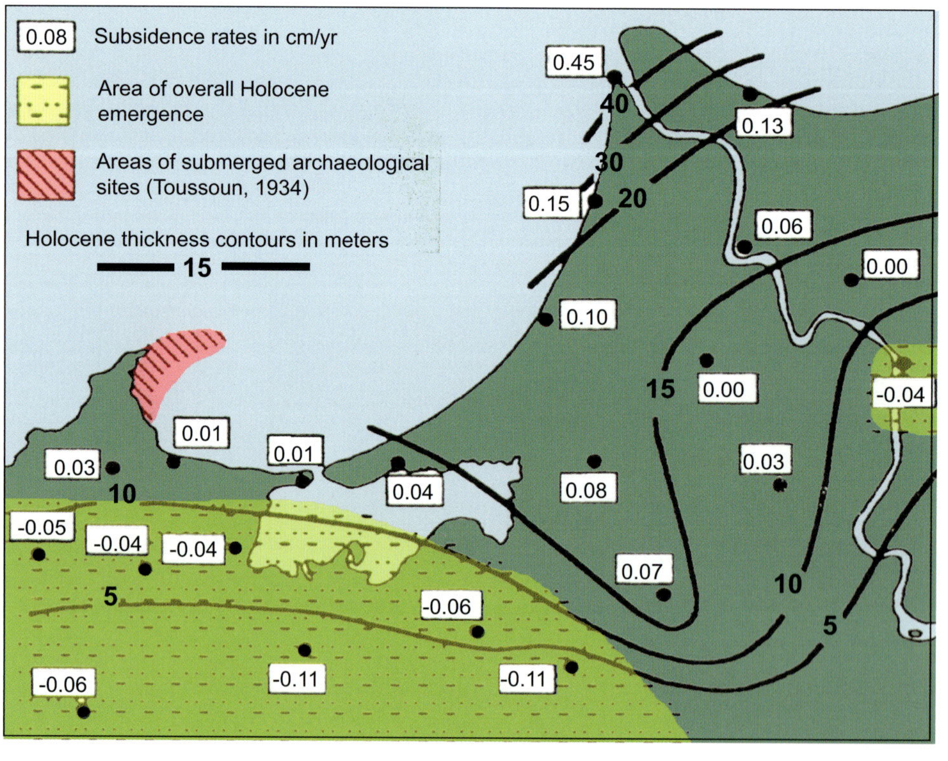

Figure 3.4 Map of Aboukir Bay margin showing the total Holocene thickness (contours in metres) and long-term average Holocene subsidence rates in cm per year. Zone of greatest subsidence now lies at the mouth of the Rosetta branch. Sector denoted by negative subsidence rates is one of overall gentle Holocene emergence (modified from Chen *et al.* 1992).

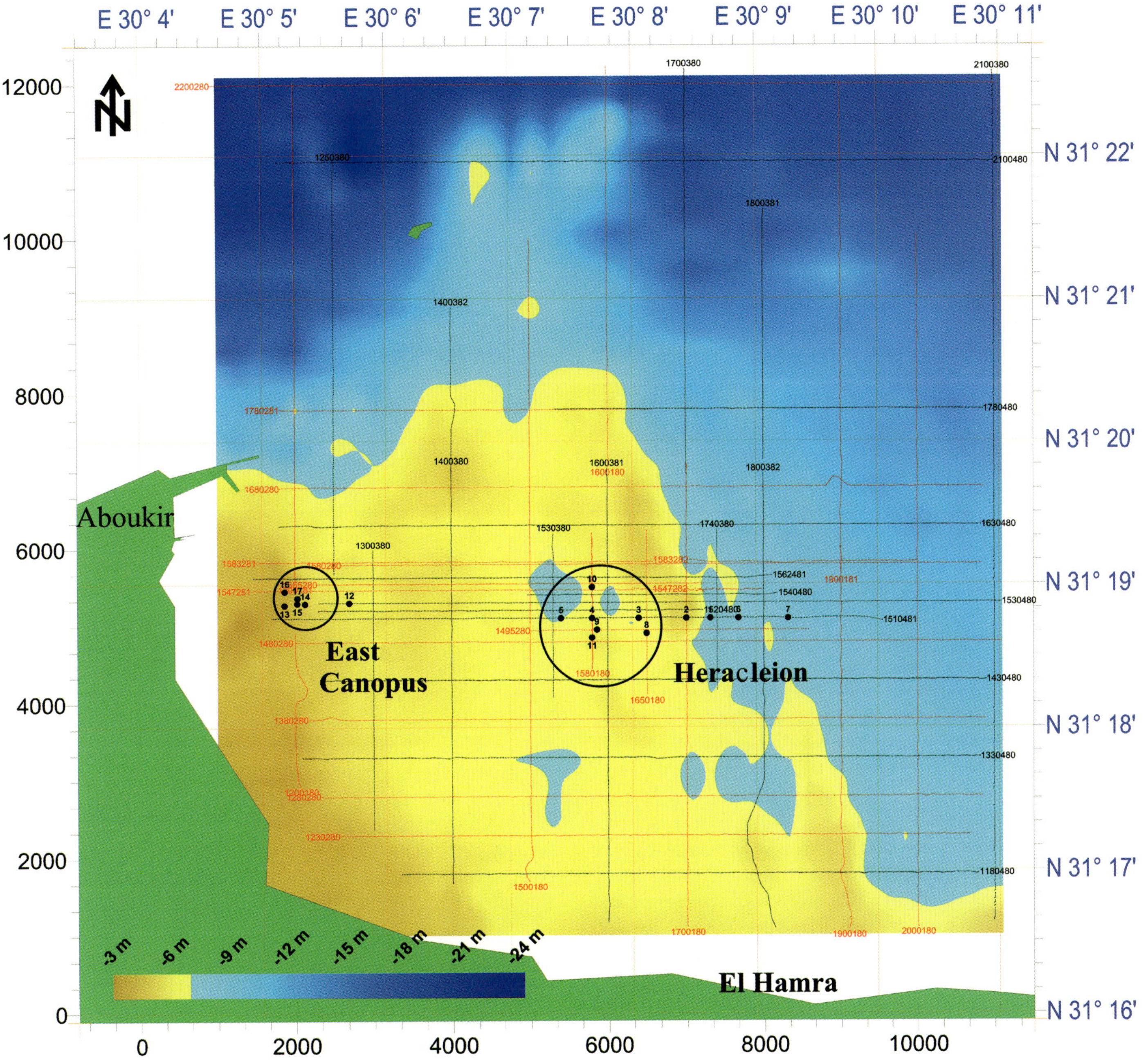

Figure 3.5 General topographical map of western Aboukir Bay showing locations of Heracleion and East Canopus and high-resolution seismic lines (selected examples of profiles are shown in Figures 3.6–3.9). Numbered dots depict the location of 17 vibracores collected in the study area. Distance scale in thousands of metres on left and bottom of figure.

presented here (Figures 3.6–3.8). The location of sediment cores, described in the following section of this chapter, are shown relative to the seismic lines and the ruins of the two submerged sites (Figure 3.5).

The east–west oriented transects are especially useful to define the configuration of the upper sediment layers (Holocene) in the western bay, including those that form the substrate directly underlying the two major submerged centres. The northernmost three profile lines, two collected north and one south of Nelson Island, reveal a highly irregular submerged topography (Figure 3.6). Higher-relief bedrock features are covered with only a thin sediment veneer, while the asymmetrically accumulated deposits at their base record the effects of deposition and scour induced by strong currents (Figure 3.9A). It is most likely that these bedrock outcrops are Pleistocene limestone of coastal origin (kurkar) of the type that forms the Aboukir peninsula and is also exposed extensively along the coastal Alexandria–Arabs Gulf region to the west (Butzer 1960; Hassouba 1980, 1995; Stanley and Hamza 1992; Warne and Stanley 1993). The two northern lines also record on the bay-floor an acoustically hard, relatively smooth, surficial sediment layer, interpreted as a thickened sequence of longshore current-swept sand. The presence of a sand cover in this sector has been verified by samples from several sedimentological surveys of Aboukir Bay (Frihy 1992a; Frihy *et al.* 1994).

The eastern segment of the fourth profile from the top of Figure 3.6 reveals a distinct channel, buried by modern sediment to a depth of ~10 ms (~7.5 m), interpreted as part of the most recent Canopic distributary system that flowed northward in what has now become the central bay. In fact, this pronounced buried Canopic channel system is recorded on most eastern segments of the seismic lines in Figures 3.6–3.8. When plotted on a map of the region, the channel can be traced almost continuously northward for >10 km from the present coast (Figure 3.10, see channel C3). The channel flowed from east of El Hamra (near El Maadiya) to an area east of Heracleion. More subtle buried channels are identified in the western part of the fourth and fifth profiles from the top of Figure 3.6, and other profiles in Figures 3.7 and 3.8. These were probably distributaries that were once associated with Canopic delta sub-lobes in the westernmost bay, and they flowed from the El Hamra region and coastal sector to the west towards the north-north-west, to Heracleion and East Canopus (Figure 3.10, channels C1 and C2).

The western segments of the east–west oriented profiles show that most Holocene and underlying Pleistocene strata of unconsolidated sediment (to depths of ~7.5 m or more below the water–sediment interface) lie above strong, well-defined reflectors. These latter are interpreted here as consolidated kurkar of Pleistocene age (Figures 3.6–3.8), comparable to the limestone exposures at and west of the Aboukir peninsula. This basal horizon was once subaerially exposed and eroded, primarily during the last glacial low sea level stand and early rise in sea level (until about 10,000–8000 years ago). In some locations, this carbonate material is faulted and forms a blocky high-and-low (horst-and-graben) configuration (see Figures 3.7, 3.8). This particular configuration pre-dates deposition of the overlying horizontally layered Holocene sediment cover (Figure 3.9C) that began to accumulate about 7500 years ago and is associated with the post-glacial rise in sea level (Chen *et al.* 1992; Stanley and Warne 1994).

Seismic profiles show that the Holocene sediment cover is about 5–7.5 m thick in the western bay, near the Aboukir peninsula. The cover thickens towards the east, and the hard basal layer (probably consolidated Pleistocene deposits) disappears at depth along the eastern segments of the profiles. Proceeding eastward of Aboukir peninsula, the Pleistocene and older units underlying the Holocene section have been lowered to more than 10 m below the water–sediment interface. This has probably occurred by downward depression, folding and/or fault offsetting of strata east of the peninsula (cf. Hassouba 1980). The phenomenon has resulted, at least in part, from the rapid accumulation of thick Nile delta deposits released from the Canopic branch until the first few centuries of the first millennium BC, and then more recently from the Rosetta branch. The effects of such a loading phenomenon, for example, are observed at the mouth of the modern Rosetta branch where the total Holocene sediment thickness increases distinctly in a seaward direction from 5 to >40 m. The downward-bowing of recent sediment sections is a function of rapid deposition, sediment compaction and weighting of the late Holocene sequence on underlying units at the river mouth (Figure 3.4).

The high-resolution seismic survey indicates that more than half of the bay study area is covered by near-horizontal stratified muds and sands that are oriented as sheets lying parallel or subparallel to the sea-floor surface (Figures 3.7, 3.8, 3.9C,E). These normally positioned strata are identified as Canopic delta sub-lobe deposits that accumulated above, and buried, the irregular (faulted, folded, eroded) pre-Holocene basal terrains. Much of the bay-floor is of low relief (<3 m), in part the result of erosion and sediment redistribution by strong bottom currents (Summerhayes *et al.* 1978; El Sayed 1979; Coleman *et al.* 1980; Inman and Jenkins 1984; Fanos 1986; Smith and Abdel-Kader 1988; Sestini 1989).

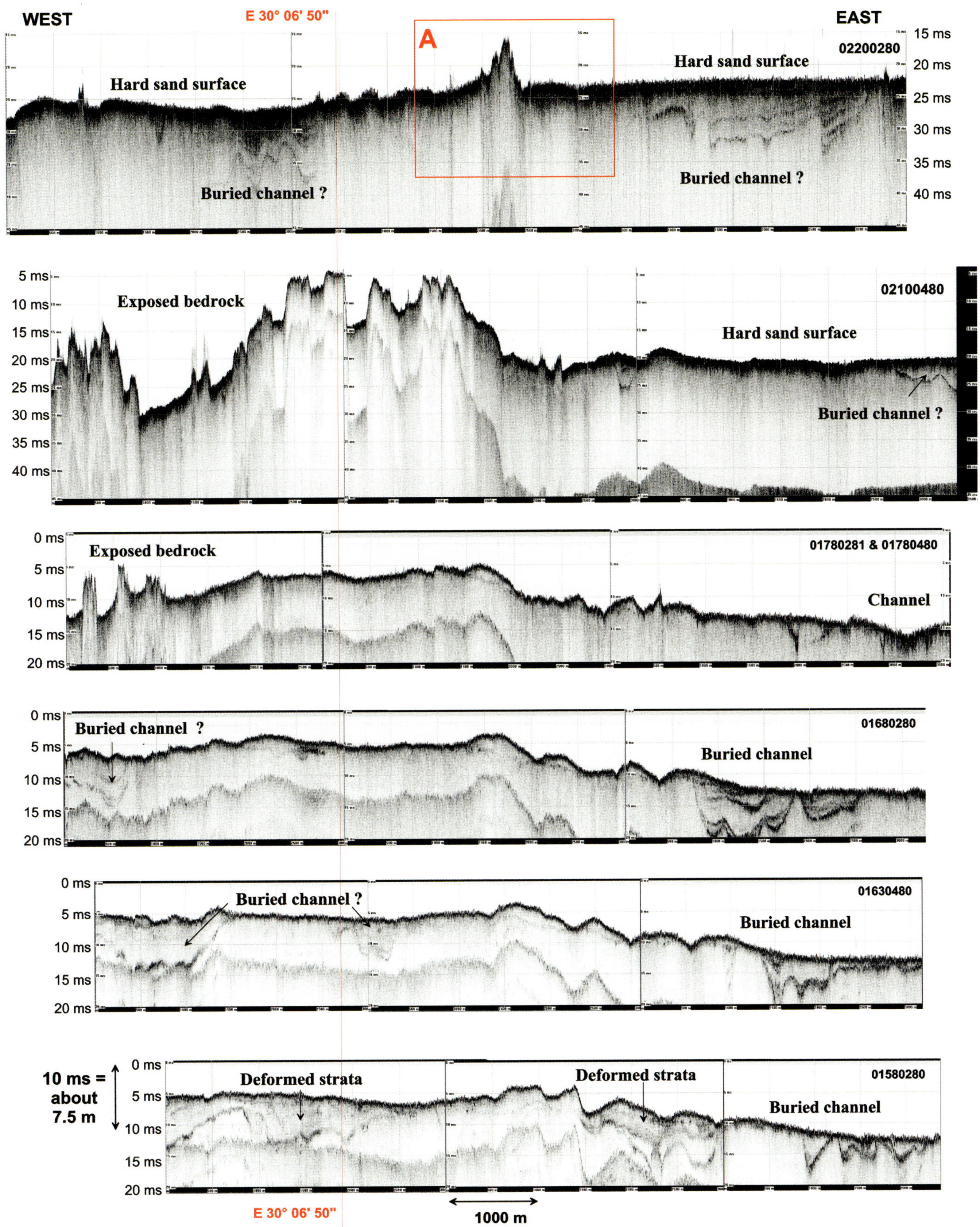

Figure 3.6 Portions of six west-to-east high-resolution seismic profiles collected in the northern part of the western bay (discussion in text). Depth in ms (10 ms = 7.5 m). Enlarged section of area A is shown and interpreted in Figure 3.9.

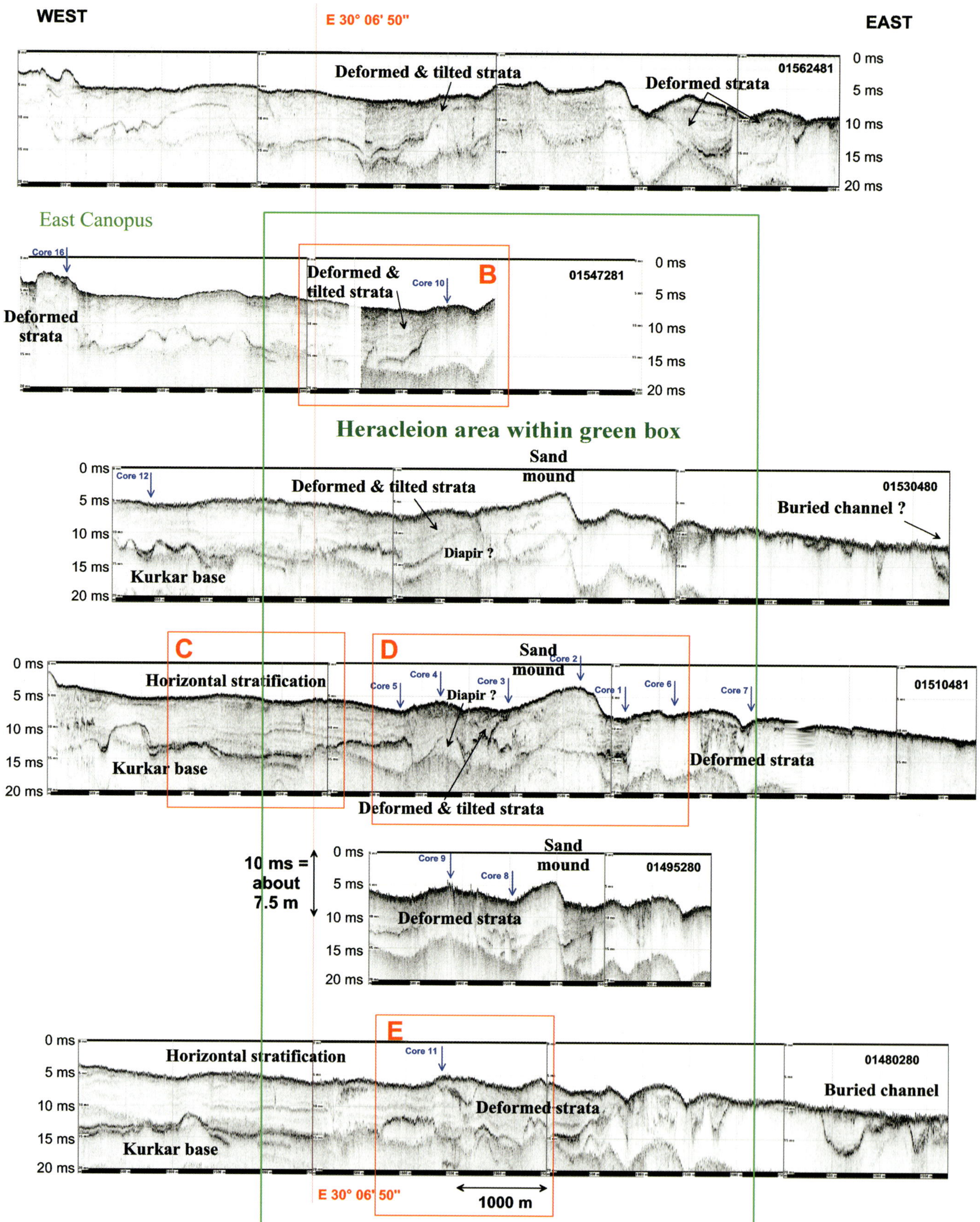

Figure 3.7 Portions of six west-to-east high-resolution seismic profiles collected in the central part of the western bay (discussion in text). Depth in ms (10 ms = 7.5 m). Enlarged sections of areas B–E are shown and interpreted in Figure 3.9.

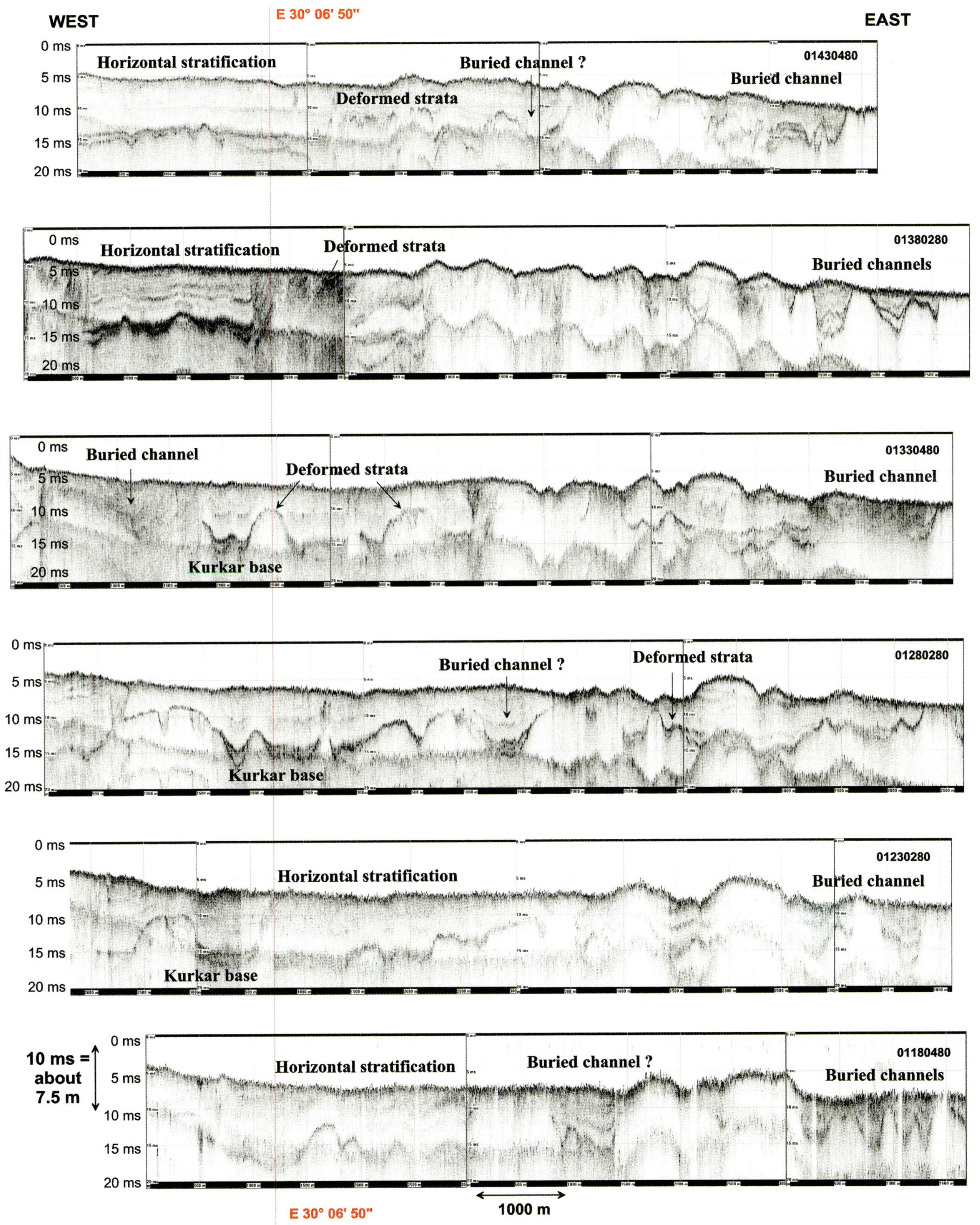

Figure 3.8 Portions of six west-to-east high-resolution seismic profiles collected in the central to southern part of the western bay (discussion in text). Depth in ms (10 ms = 7.5 m).

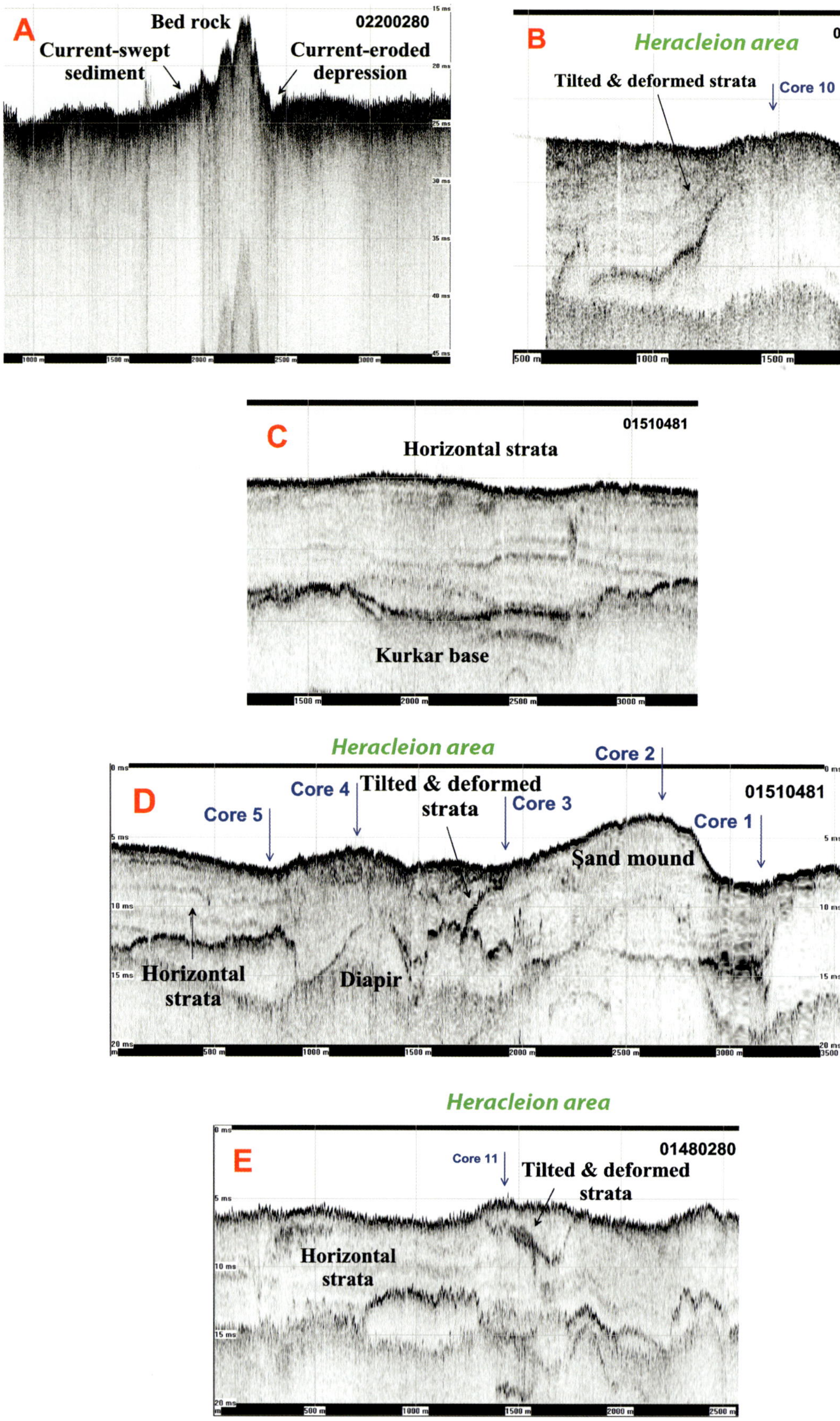

Figure 3.9 Enlarged sections of high-resolution profiles from Figures 3.6 and 3.7. Disturbed (tilted and deformed) Holocene strata prevail under Heracleion (B, D, E), while horizontal sediment layers in bay sectors away from the ruins of the two cities. Interpretation of features are shown on the profiles and discussed in the text. Depth in ms (10 ms = 7.5 m).

In sharp contrast to the above finding, seismic profiles show that the sedimentary cover is not flat-lying in some well-defined and aerially restricted sectors of the bay. Two such areas, interestingly, are those specifically underlying Heracleion and East Canopus. One anomalous surficial feature is the large domed horizon located just to the east of the submerged Heracleion ruins (Figure 3.7, third to fifth profiles from the top of the figure). This feature is nearly 800 m wide and 4 m high, with its steeper face sloping to the east (Figure 3.9D). Core 2, collected on this mounded topography, indicates that the upper part of the feature (at least to the core base at a depth of 3.2 m) is formed of sand (log of this boring shown in Figure 3.17). In cross-section, the configuration of the asymmetrically shaped mound resembles a broad, low-lying dune; it is bounded on its eastern margin by a 250–300 m wide channel-like depression, in which core 1 was recovered (Figure 3.9D). It is actually quite remarkable that this relief feature, formed of unconsolidated sand, has not been eroded and obliterated by the bottom currents that sweep the bay-floor. In fact, the set of east–west oriented profiles crossing over it at several places (Figure 3.7) indicates that this feature is an elongate north–south oriented sand bank. The sands appear associated in some manner with the offset strata that occur immediately to the west of it, and which form the highly disturbed sediment substrate lying immediately beneath Heracleion (Figure 3.9D). Although the origin of this elongate relief feature has not yet been determined, it is possible that it is a partially preserved coastal sand ridge that once lay parallel to one of the Canopic sub-lobe channels to the east of Heracleion. As such, it may have originated by the reworking of natural levee sand by fluvial and marine coastal processes. Another interpretation suggests an accretionary sand deposit related to some large presently buried feature, natural or anthropogenic, now lying at depth below the surface. In the latter case, one could envision that a relief feature such as a wall may once have extended above the sea-floor surface; such a structure would have trapped sand driven towards the east, first by winds and then, after submergence, by the moderate to strong easterly driven currents (to 60 cm/sec) flowing along the Nile delta coast.

Of special significance with regard to the present study are deformed strata that lie directly beneath the sites of both Heracleion (Figure 3.9B,D,E) and East Canopus (Figure 3.7, see western sector of the second profile from the top of the figure, near core 16). From this regional seismic survey, we observe that patches of strata beneath the ruins at both sites in the western bay are not horizontal, but markedly disturbed, tilted and uplifted. Some possible mound-like diapirs (post-depositional, squeezed upward, domed sediment features) are also observed at depth (Figure 3.9D). Originally, these deformed Holocene horizons were laid down horizontally at and seaward of the delta mouths. From the profiles, we now have a record of geologically recent syn- and post-depositional events that have raised and offset some of the deeper (including early to mid-Holocene) sediment layers that once lay beneath the two cities all the way up to the present sea-floor surface (Figure 3.9B,D,E).

Side-scan sonar images collected by the IEASM in the western bay provide some additional evidence of anomalous stratification presently visible at the bay-floor surface, especially in the vicinity of and north of Heracleion (close to the seismic profile segment shown in Figure 3.9D). For example, about 200 m north-east of the submerged ruins that lie north of Heracleion (Figure 3.11, lower image), we find an anomalous but well-defined feature resembling undulations of a pleated fabric. An enlarged image of this feature (Figure 3.11, upper panel) shows that it occupies a restricted, well-defined area (diameter ~150 m). We suggest tentatively that the undulations comprise partially exposed strata (probably tilted and mud-rich) that were uplifted to the bay-floor in fairly recent times. These were subsequently modified by bottom currents, and are now covered by a thin veneer of modern rippled marine sand.

A precise nuclear resonance magnetometer survey made by the IEASM across the same area (Figure 3.12A) also records distinct anomalies probably relating to offset Holocene stratification at and just beneath the sea-floor surface in the vicinity of the two submerged centres. Well-defined straight and curvilinear features are mapped at both localities, some trending north–south at East Canopus and ENE–WSW at Heracleion (Figure 3.12B,C). During the archaeological reconnaissance done by the IEASM, based on the geophysics survey, divers reported that these anomalies do not appear to coincide with any obvious features, geological or archaeological, exposed at the bay-floor surface. However, IEASM archaeological excavation of one such anomaly discovered near the temple at East Canopus showed it to be a long (~100 m or longer), buried, sand-filled trench naturally formed in the underlying Holocene mud substrate (Figure 3.12B, T = trench). Cleared of sand, the curvilinear trench is V-shaped in cross-section, about 5 m wide at the top and 2 m deep from the bay-floor surface to its base (Figure 3.13A,C); a well-defined, rectilinear fault-like break (to ~50 cm wide and deep) is present in the mud that forms its base (Figure 3.13A,B).

The subsequent analysis performed by IEASM (Goddio F, 2007) indicate that the trench was artifi-

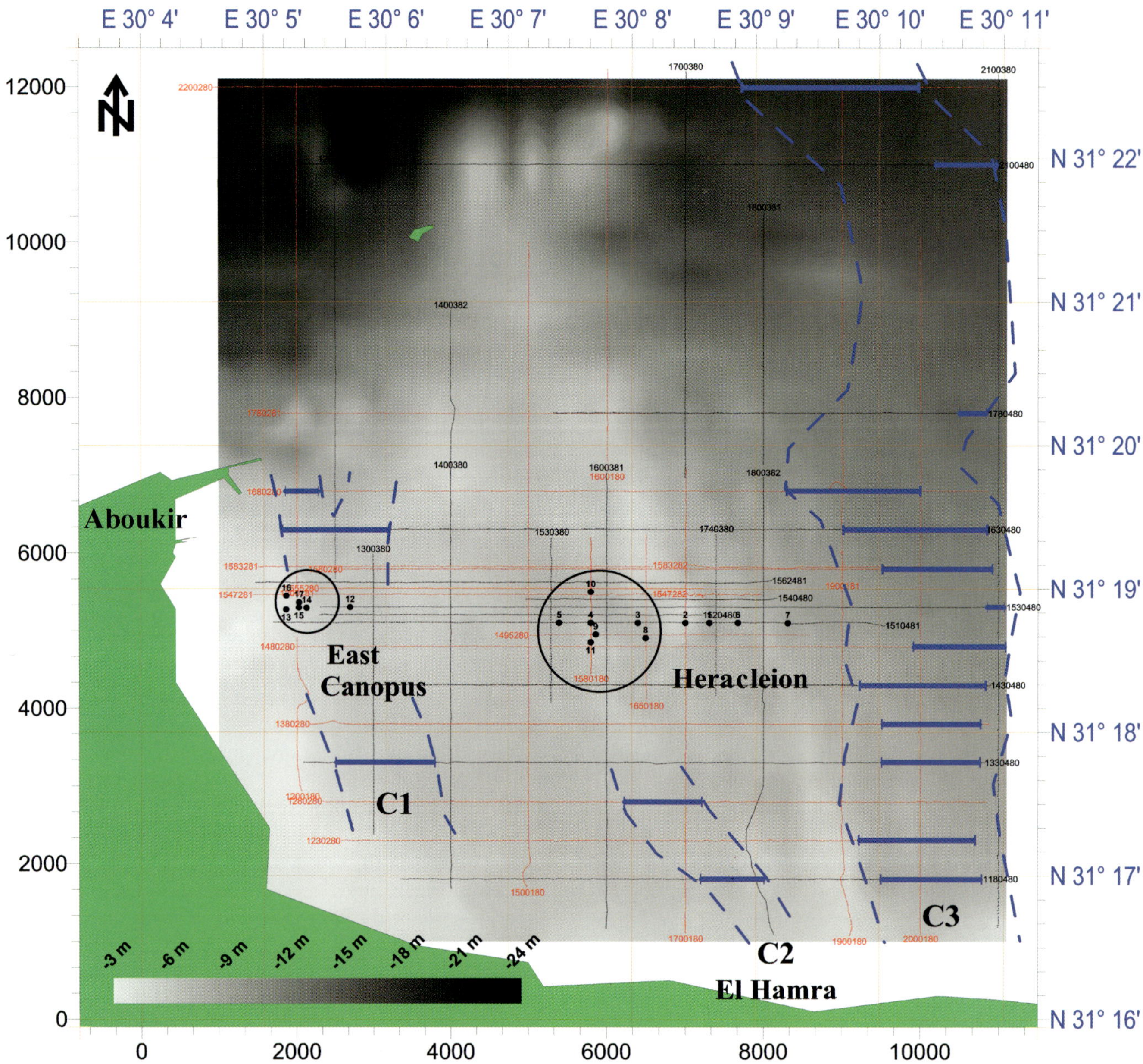

Figure 3.10 Configuration of three Canopic sub-lobe channels (C1, C2, C3) that once extended from the present coast towards the north, based on data in high-resolution seismic profiles (Figures 3.6–3.8). Note that the two older channels (C1, C2) that once flowed to Heracleion and East Canopus have also been mapped in geological studies on the adjacent coastal margin to the south (Chen *et al.* 1992; Warne and Stanley 1993). The younger channel (C3) extending from east of El Hamra has also been detected in earlier investigations (El Bouseily and Frihy 1984; El Fattah and Frihy 1988). *Courtesy IEASM*

Facing page: **Figure 3.11** Side-scan sonar images obtained by temple in the Heracleion area. Lower panel shows the configuration of the archaeological ruins of that city now under water at depths of 5–7 m. Upper panel is an enlarged portion of the region north-east of Heracleion, showing ruins and a series of wave-like ridges interpreted as tilted and uplifted Holocene strata exposed at the sea-floor (Figures 3.7, 3.9). The offset strata have subsequently been reworked by bottom currents and covered with a rippled sand veneer. Each side-scan swath equals 80 m. *Courtesy IEASM*

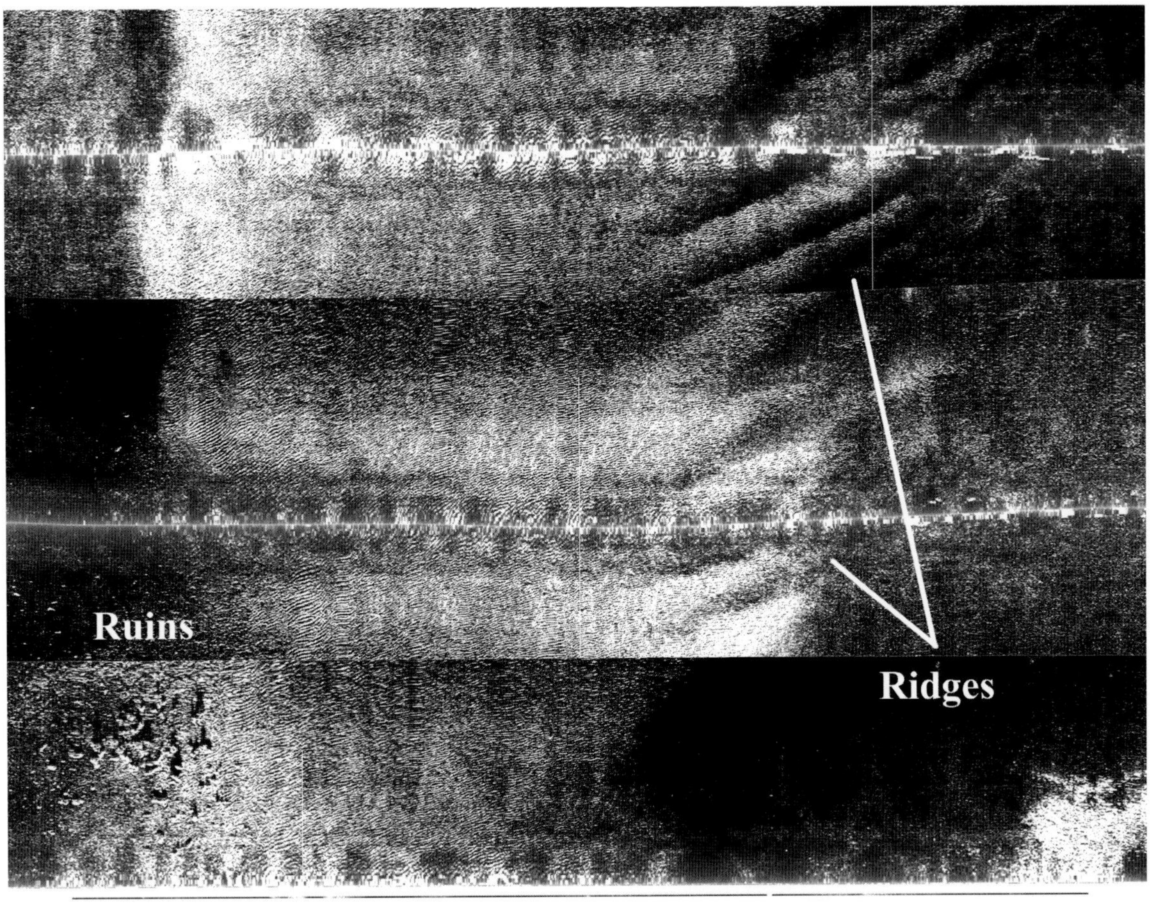
Ruins
Ridges

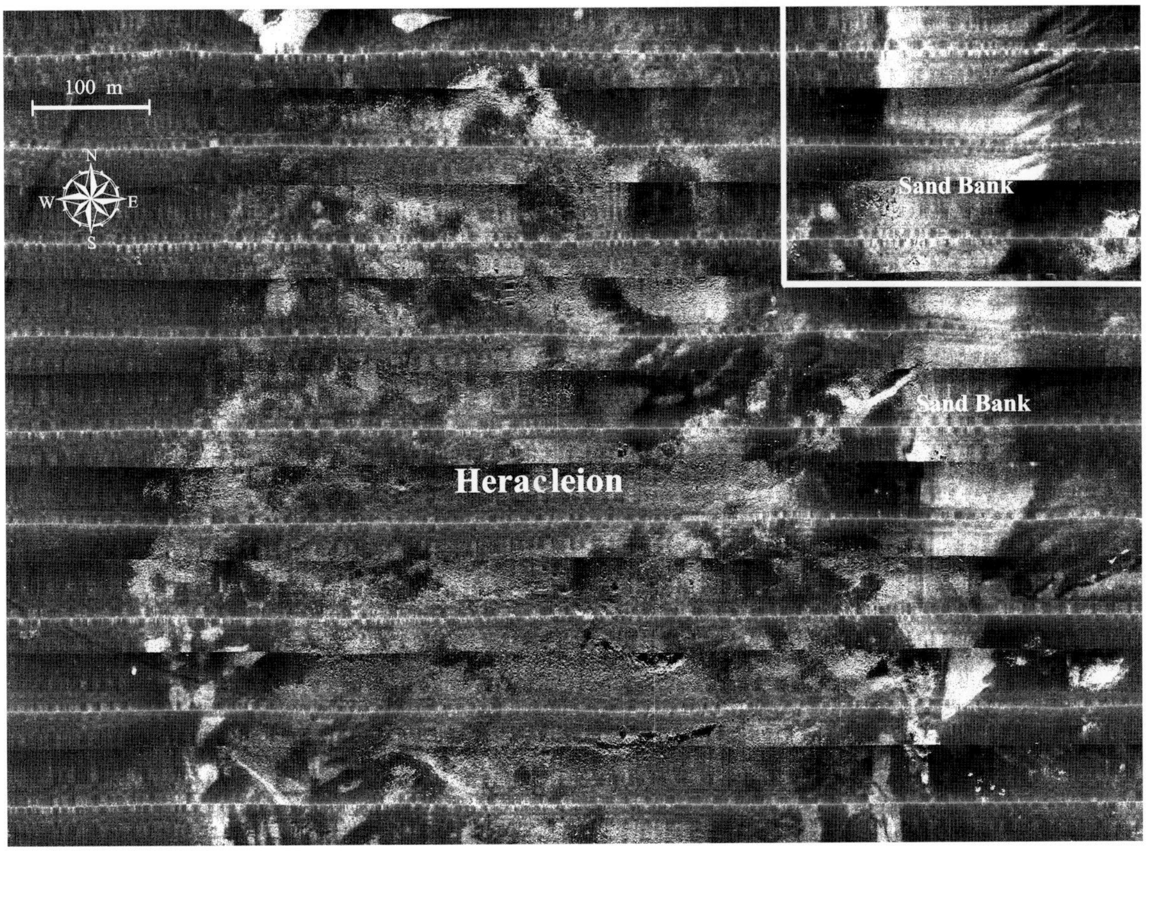
100 m
N
W
E
S
Sand Bank
Sand Bank
Heracleion

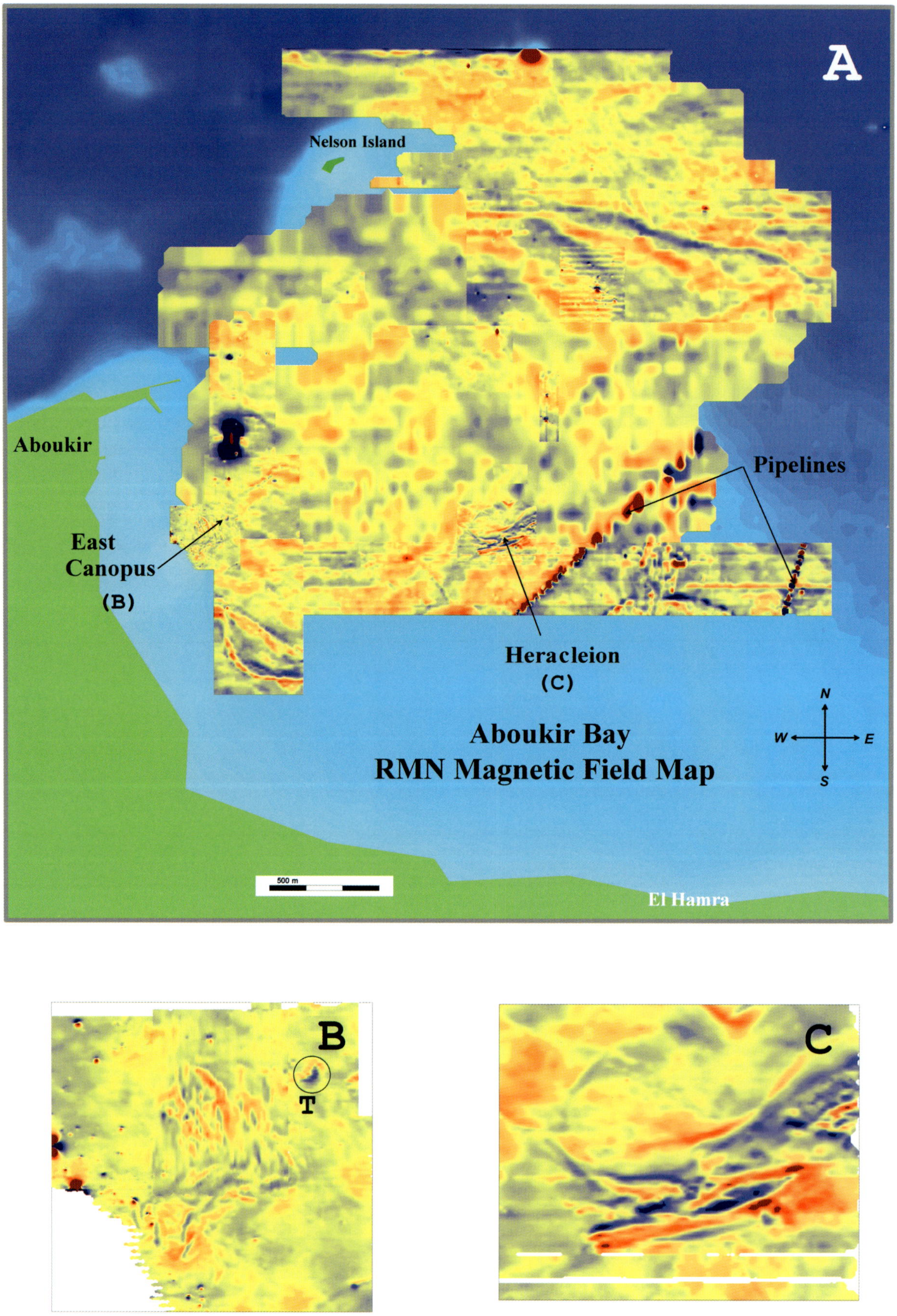
A
Nelson Island
Aboukir
East
Canopus
(B)
Pipelines
Heracleion
(C)
Aboukir Bay
RMN Magnetic Field Map
N
W
E
S
500 m
El Hamra
B
T
C

cially filled by humans at a time when the adjacent man-built structures were still positioned at an elevation above sea level (Figure 3.14): the sand fill is very well sorted quartz-rich sand (mode, 0.223 mm) that is almost identical in texture and composition to modern beach sand (mode, 0.223 mm) and modern dune sand (mode, 0.203 mm) of the type collected in spring 2000 along the present adjacent Aboukir coast. The trench fill sand, on the other hand, differs from the somewhat coarse modern marine sand that covers the present bay-floor surface: the mode of the latter is somewhat coarser (0.269 mm), and its composition includes a higher proportion of marine shell debris than the artificial trench fill sand.

Even more obvious evidence that the large V-shaped trench at East Canopus had formed and was still above sea level while East Canopus was inhabited is the abundant plant matter found along the trench walls (Figure 3.13C). Diver observations show the trench walls had been originally lined by mats of fresh to brackish water plant material, principally phragmites (preliminary observation determined by Scanning Electron Microscopy). These mats were placed on the walls before the depression was filled with well-sorted beach and dune sand. These plant mats have left undulations on the trench walls (Figure 3.13A,C). Additional observations supporting the argument for the original subaerial position of the trench are the presence in the mud base of well-preserved, cloven-hoof bovid (probably cow) tracks (Figure 3.13D,E) and the bones of an antelope (Figure 3.13F). The bone of the latter has been radiocarbon dated to the first century BC. Of the distinct magnetic anomalies noted on the magnetometer map in Figure 3.12, only the excavated feature we describe here at East Canopus has been studied to date.

5 Disturbed stratification and anomalous radiocarbon dates in surficial sediment

A suite of sediment cores was collected at and between the archaeological sites in the study area to provide a more thorough investigation of the processes resulting in site submergence. This database supplements information from coastal and sediment borings recovered previously in the western bay area (Attia 1954; Frihy 1992a; Stanley *et al.* 1996). The new set of cores includes a series of 17 vibracores (Figure 3.15), ranging in length from 1.45 m to 5.51 m and penetrating only to the upper to middle parts of the Holocene sediment sequence. The continuous sections were collected in 10 cm diameter aluminium tubes with Rossfelder unit P-3 equipment in April 2001 (Figures 3.16). Seven of the core localities were at and near Heracleion (cores 3A–5A and 8A–11A), four were sampled at and near East Canopus (cores 14A–17A), and six were recovered in bay sectors away from the two submerged sites (cores 1A, 2A, 6A and 7A east of Heracleion, and cores 12A and 13A between Heracleion and East Canopus).

The simplified lithologic logs of the 17 cores show the presence of a thin (a few cm to <1.0 m), modern marine sand layer at most core tops (Figure 3.17). This marine sand comprises a mix of terrigenous components of Nile derivation, carbonate particles (some transported eastward from the region west of the Aboukir peninsula) and shell material (molluscs and other organisms). The surface marine sand covers diverse lithologies, best observed in split core sections and in X-radiographs (Figure 3.18). These include: dark organic-rich mud (silt and clay components), appearing uniform and non-laminated (Figure 3.18A), or mud formed of well-laminated alternating silt and clay (Figure 3.18B), or strata composed of fine-grained sand with interbedded whole and broken mollusc shells (Figure 3.18C); sand, for the most part moderately to well sorted and of fine to medium grain size; and, in 6 of the 17 borings, contorted units that display soft sediment deformation (Figure 3.18D,E).

Contorted strata are formed of mud, or sand, or mixes thereof. One of these deformed silt and clay layers includes a ceramic artefact and a kurkar rock fragment (Figure 3.18D). Soft sediment deformation is exemplified by vertical sediment flow (Figure 3.18D) and flame structures (Figure 3.18E, arrow), typical of fluidisation (where loose, water-saturated material has flowed in a liquid-like fashion). The seven sections of disturbed sediment in the six borings range from about 30 to 60 cm in thickness, and these occur primarily in the upper 3 m of section in cores 3B, 5B, 7A and 9A recovered at and near Heracleion, and in cores 15A and 16A at East Canopus. Moreover, it is also noted that cores recovered at or close to the two submerged archaeological ruins display a larger proportion of sand than those collected at some

Facing page: **Figure 3.12** (A) Results of the IEASM nuclear resonance magnetometer survey across the study area. Compare natural anomalies with the hard, well-defined signals produced by pipelines in the south-east study area. (B) Anomalies, including the diver-excavated trench (T), at East Canopus (details in Figure 3.13); trends in this area are oriented north–south and east–west. (C) Anomalies, better defined, in the Heracleion region; trends here are oriented ENE-WSW. Horizontal scale in A. *Courtesy IEASM*

A
B
C
(P)
D
E
F

distance from the two sites (for example core 12A formed only of mud). Only core 2A, collected in the broad elongate mound east of Heracleion (discussed in the previous section), and core 6A, east of the mound, were comprised entirely of sand.

Radiocarbon dates, determined primarily by AMS methodology and using mainly plant matter for the analysis, were recorded for 41 vibracore samples (analyses by Beta Analytic Inc., Miami, Florida). The dates listed here are in uncalibrated radiocarbon ages (Figure 3.17). Unusually old dates, ranging from >5000 to 6880±60 years before present (BP), were obtained for core samples near the core tops at both archaeological centres, and in two borings where the upper marine sand layer was missing (cores 1A and 12A). In contrast, much younger dates (to 1980±40 yr BP) were obtained in the upper parts of some cores collected in the bay away from the ruins (see, for example, core 13A).

In an effort better to comprehend the reasons for this very broad range of dates in the upper part of the Holocene core sections, an additional set of 12 radiocarbon dates was obtained from surface and near-surface localities at East Canopus. Samples of peat and organic-rich sediment were collected by divers beneath archaeological structures at diverse sites; recovered sediment layers were exposed close to the bay-floor surface, just below the thin, uppermost, marine sand cover. It was anticipated that the radiocarbon analyses of near-surface samples might provide the age of the sediment substrate on which this site was directly built. Of the 12 near-surface

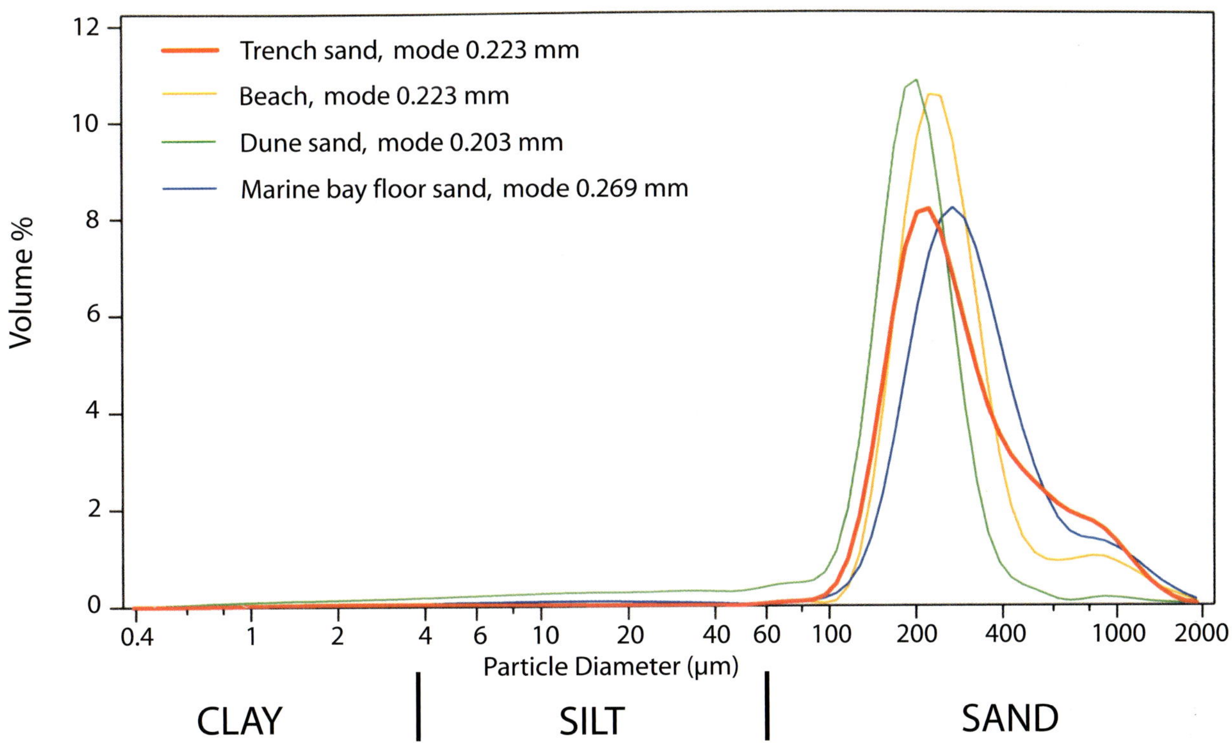

Figure 3.14 Grain size distribution of sand fill collected in the diver-excavated trench at East Canopus (red size curve). This material, placed in the trench by man, is similar in size and sorting to modern beach (yellow curve) and dune (green curve) sands collected along the present adjacent coast, but is finer grained than the marine sand (blue) that floors much of the bay and covered the top of the trench before excavation.

Facing page: **Figure 3.13** Photographs collected in the IEASM archaeological excavation of the trench (shown as T in Figure 3.12B). (A, B) Section of trench after removal of artificial sand fill; note rectilinear fault-like feature in the mud base of the trench (~50 cm wide and deep). (C) Part of the exposed trench wall with abundant plant matter (P, arrows); the trench wall was lined with mats of brackish wetland flora, and these caused undulations observed here and in A. (D, E) Cloven-hoof bovid (probably cow) tracks preserved in mud at bottom of trench. (F) Bones of antelope, radiocarbon date: 40BC +/-45 years. *Courtesy IEASM*

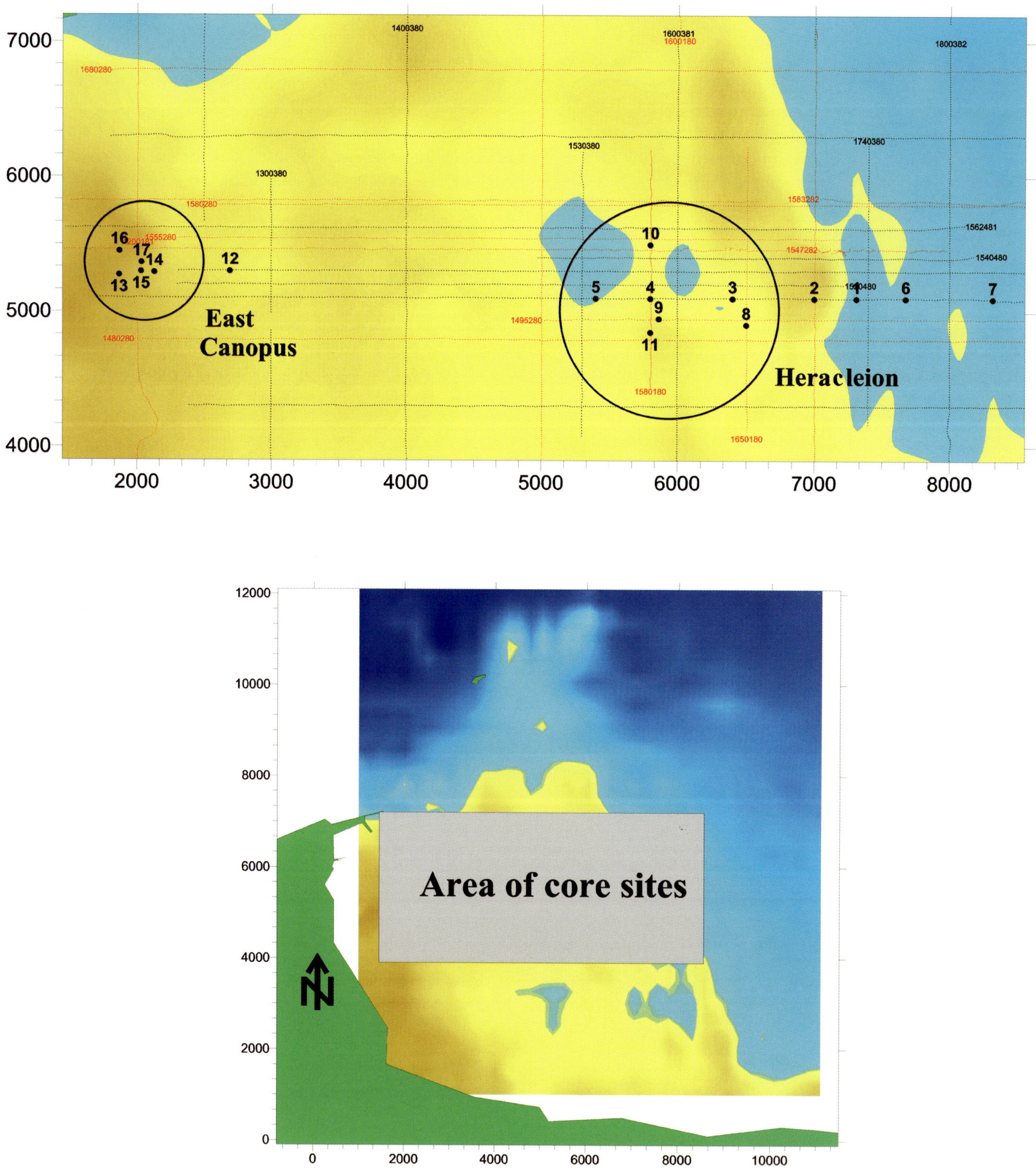

Figure 3.15 Map of western Aboukir Bay showing locations of high-resolution seismic lines and 17 vibracore sites in areas near and between Heracleion and East Canopus.

Facing page: **Figure 3.16** Photographs showing: (A) vibracore drilling; (B) recovery of a sediment-filled aluminium core section; and (C) splitting of an initial test core aboard the boat, with expedition leader Franck Goddio in foreground (April 2001).

A
63-97
B
C

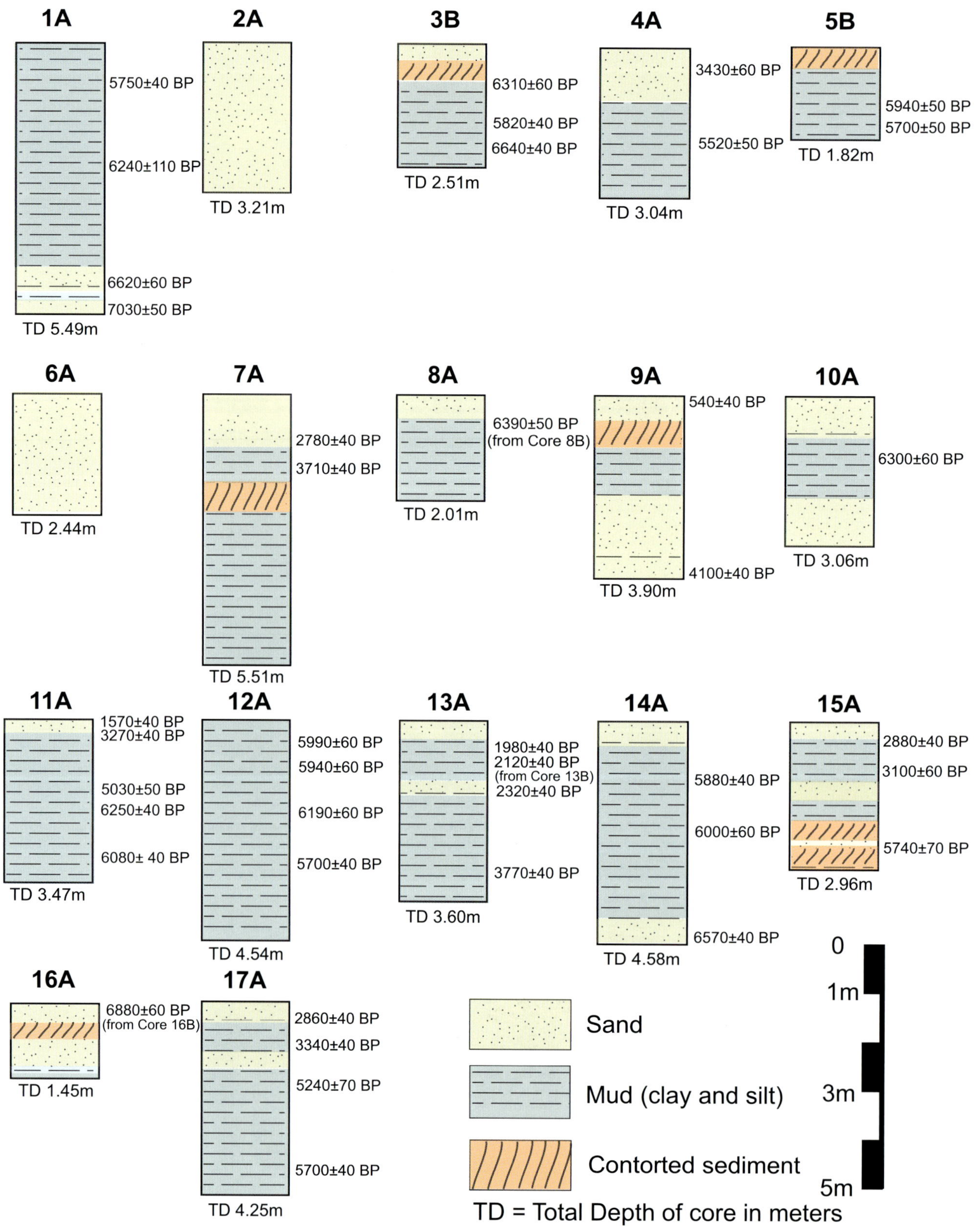

Figure 3.17 Simplified lithologic logs of 17 vibracores collected in western Aboukir Bay. Dates are in years before present (uncalibrated radiocarbon ages). Note the seven horizons of contorted (fluidised) sediment sections in six of the cores, all collected at or near the archaeological sites.

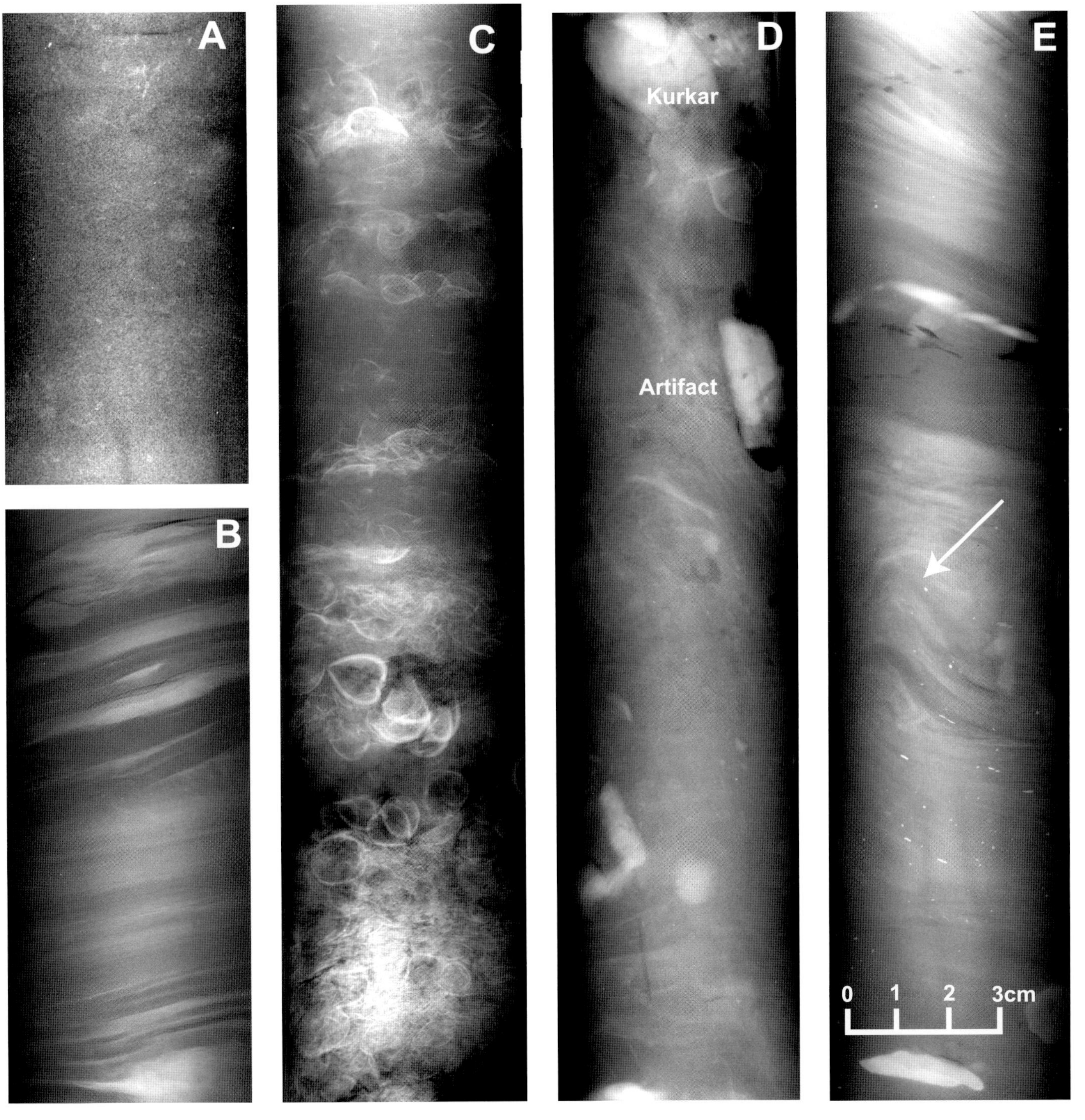

Figure 3.18 Selected X-radiographs of split-core sections from the study area showing various sedimentary features discussed in text (depths in cm from core top). (A) Core 12, massive structureless organic mud (5–17 cm). (B) Core 13A, inclined laminated silts (light coloured layers) and clays (6–22 cm). (C) Core 15, horizontally bedded silty mud with shell horizons (95–122 cm). (D) Core 5B, with kurkar and artefact fragment in a fluidised mud section (0–27 cm). (E) Core 9, fluidised section with well-developed flame structure shown by arrow (87–114 cm).

samples, three were dated at ~2000 yr BP and five from 2250 to 2360 yr BP, and these eight results appear 'reasonable' in that they fall within an expected age range for the archaeological site substrate. However, we must also take note of 4 of the 12 other surface samples: there was one older date of 3130 yr BP, and three much older dates, ranging from 5620 to 6750 yr BP. Thus, in this analysis, we find that radiocarbon-dating of both diver-recovered surficial samples and material from the upper parts of cores provides similar results, i.e. a mix of dates at the bay-floor that can vary by as much as 5000 years. Overall, the radiocarbon analyses indicate the presence of unexpectedly 'old age' sediment sections (mid-to-late Holocene) in two settings: (1) just below man-built structures and cultural horizons at both sites, in sectors where sediment appears to have been disturbed (tilted, fluidised); and (2) away from the archaeological sites, where the upper mid-to-late Holocene sections may have been eroded (probably removed by bottom currents), such as at core localities 1A and 12A.

Still another unexpected observation was made with respect to the faunal content in the 12 diver-collected surface samples at East Canopus, all of which were recovered from what is now a fully marine setting. Preliminary analyses of the molluscs and microfauna (ostracods, foraminifera) in these samples record only a modest proportion of eastern Mediterranean open-marine fauna. Rather, many samples are dominated by a mix of brackish with some freshwater species, and the presence of only a relatively small proportion of marine forms (Drs M.P. Bernasconi, N. Pugliese and R. Mellis, 2001, personal communication and 2006, Chapter 4, in this volume). The significance of the above assemblage of geological, sedimentological and faunal findings with respect to site submergence is evaluated in following sections.

6 Submergence by gradual processes: causes and effects

It is proposed here that the two archaeological sites in Aboukir Bay have been lowered, at least in part, by gradual long-term natural processes (Figure 3.24, emphasised by Category A). The two major concurrent processes active in the study area during the past 25 centuries are a progressive rise in sea level and a concurrent land subsidence. These are phenomena recognised as having affected many of the world's continental margins to varying degrees during the Holocene to the present (Mörner 1971; Pirazzoli 1992).

During the past several millennia, much of the sea level rise that has affected oceans on a worldwide basis (termed eustatic change) is the result of polar ice melt and thermal expansion of ocean waters (Fairbanks 1989). The rate of rise has been measured in the Mediterranean by various workers who provide values that have ranged from less than, to little more than, 1.0 mm per year since the mid-Holocene (Pirazzoli 1987, 1992; Milliman and Haq 1996). Also contributing to sea level rise effects is the contemporaneous lowering of land at the Nile delta margin, owing to compaction of the underlying water-saturated deposits in the bay, isostatic depression of land mass in the north-western Nile delta, and regional changes in the geoid through time. In the case of the western Aboukir Bay area, measurements indicate that the substrate over the long term has been lowered at a rate of about 1 mm per year, excluding the effect of eustatic rise in sea level (Chen *et al.* 1992; Warne and Stanley 1995). Thus, the cumulative effects of eustatic rise and land subsidence, together, account for a maximum value of about 2.0 mm per year of overall sea level rise (this compares to >2.5 mm per year for the Nile delta margin as a whole: Stanley and Warne 1993a,b, 1998; Sestini 1989). The combined result of these two factors, resulting in a rise in sea level, is termed relative sea level rise.

The calculated values of relative sea level rise can be used to determine the gradual effects through time at both Heracleion and East Canopus in the western bay. For calculations here, we consider that: (1) the large human-built structures at the sites were originally emplaced at 2 m above sea level (most of the adjacent delta wetlands surrounding the settlements were at much lower elevations (<1 m above mean sea level; cf. Kerambrun 1986; Loizeau and Stanley 1993); and (2) the submerged ruins presently lie on the bay-floor at a depth of 5–7 m (for calculation, an average value of 6 m is used). This accounts for a total lowering of ~8 m. To measure the rate of subsidence also requires that the length of time since the cities were built be incorporated in the calculation. For this, it is assumed that initial construction began after the pharaoh Psamtik (Psammetichos) I authorised Greek merchants to trade in Egypt in the seventh century BC. It is probable that settlement at the two coastal cities was well under way by the sixth century BC, during the reign of Amasis, when the town of Naukratis, a Greek concession on Egyptian soil, had developed into a major commercial centre (Coulson and Leonard 1979; Lévêque 1994). Thus, the total time span between the period of construction and present rediscovery of the ruins is approximately 2500 years.

About 12 centuries had elapsed between the period of construction at 2 m above sea level in the sixth century BC and the time when Sophronios of Jerusalem described the temple of the Evangelists as still standing at the shoreline at the beginning of the seventh century BC. This indicates that relative sea level rise was about 2 m in 1200 years, indicating a minimal mean rate of rise of 1.7 mm per year (Figure 3.24, Category A). A roughly comparable rate of annual relative sea level rise (2.0 mm/yr) for this region has been previously calculated independently for the mid-to-late Holocene period, including the time span from 700 BC to the present (Chen *et al.* 1992; Warne and Stanley 1993, 1995). By applying a value of 2.0 mm/yr during the past 13 centuries, from Byzantine time to 2000 BC, an additional relative rise of sea level of 2.6 m is derived. This would thus account for a total rise of 4.6 m in 2500 years, i.e. somewhat more than half of the total change in lowered elevation since the original construction of structures at Heracleion (Figure 3.24). By itself, this gradual rise would have caused the submergence of large areas of low-lying delta margin and considerable retreat of the shoreline in a landward, largely southward, direction (El Fishawi and Fanos 1989).

A long-term average rate of 2.0 mm/yr of relative sea level rise is a rather modest one for the coastal margin of a large delta (Pirazzoli 1992; Milliman and Haq 1996). Nevertheless, the cumulative effect and damage that this gradual process produced on the Aboukir Bay settlements over several centuries would have been remarkable. To illustrate this phenomenon, it is useful to consider the effects of such relative sea level rise along the coast of a well-studied delta, that of the Mississippi River on the Gulf of Mexico margin. Examination of such a large modern delta provides insights on processes affecting the earlier Canopic delta sub-lobes in the study area. The Nile and Mississippi systems are characterised by some similar attributes, such as flow of their respective river channels into shallow water bodies (bays) affected by tides of low range (~30 cm in the Mediterranean, with a mean range of 12 cm and maximum range of 59 cm in Aboukir Bay; Sestini 1989). On the other hand, the pre-Aswan Nile accounted for a much lower water discharge and sediment load carried to the coast than the larger Mississippi River, thus resulting in smaller Canopic branch delta sub-lobe construction at a somewhat slower rate. Moreover, the Mississippi delta sub-lobe margins are affected by destructive hurricane surges (usually in late summer and fall) and high rates of submergence, while those of the Canopic Nile were eroded by powerful winter storm waves (to >2 m in height; Sestini 1989; Nafas *et al.* 1991) and generally stronger coastal currents (Smith and Abdel-Kader 1988). In the Mississippi delta, natural processes have caused a higher overall rate of wetland loss, resulting in extraordinarily rapid marsh erosion (wetland loss every hour is estimated to be roughly equivalent to the size of a football field). Examination of this delta margin loss and wetland submergence in the lower Mississippi delta (Penland *et al.* 1988) is informative for the insight these provide on the disappearance of the two archaeological sites in Aboukir Bay.

From the early 1800s to the present, for example, the Mississippi deltaic coastal setting has undergone a repetitive cyclic sequence of construction, growth and progradation, followed by destruction and submergence. At least six delta sub-lobes have evolved near the mouth of the Mississippi River in less than two centuries (Figure 3.19, upper panel). To illustrate what may have happened in Aboukir Bay when the Canopic branch was active, we review here the evolution of one of these modern Mississippi sub-lobes, such as the one formed near the mouth of a major Mississippi distributary in West Bay (Figure 3.19, lower panel). The Mississippi river breached its western natural levee at flood stage in the 1830s at a site termed The Jump, and in the years that followed tons of fine sand, silt and clay were carried through this crevasse into shallow West Bay. By the 1870s, only 40 or so years later, a new series of well-formed, radiating sand-silt channels, levees and channel mouth bars had developed. Several of the channels and their low-lying adjoining marshes reached maximum progradational growth phase by the 1920s. The progradational build-out of the new sub-lobe subsequently began to slow and, by the 1950s, was discontinued principally as the result of a gradual, long-term relative sea level rise. Sea level rise resulted from progressive eustatic elevation effects and lowering of the bay-floor by compaction of underlying sediment and isostatic depression. As these latter processes continued during the past 50 years, the sub-lobe channel elevation decreased, progressively more marsh was lost, and shallow lakes developed in areas of former wetlands. Consequently, marshes and associated channels of the West Bay sub-lobe have been disappearing below the waves during a natural destruction phase. Many other delta sub-lobes have undergone an initiation–construction–destruction evolution similar to the one in West Bay, and these cycles in the lower Mississippi occur during periods lasting little more than a century (Elliott 1986; Coleman 1988; Penland *et al.* 1988).

We propose that similar cycles of sub-lobe development involving initiation, construction, build-out and destruction occurred at the mouth of the Canopic branch (Stanley and Warne, Chapter 2 in this volume), but at a slower rate than at the mouth of the larger and much more powerful Mississippi. Evidence for such a shifting lobe and delta margin evolution has been provided by previous studies of Aboukir Bay (Chen *et al.* 1992). Criteria include the presence of at least three major Canopic channels since the mid-Holocene: two older channels were

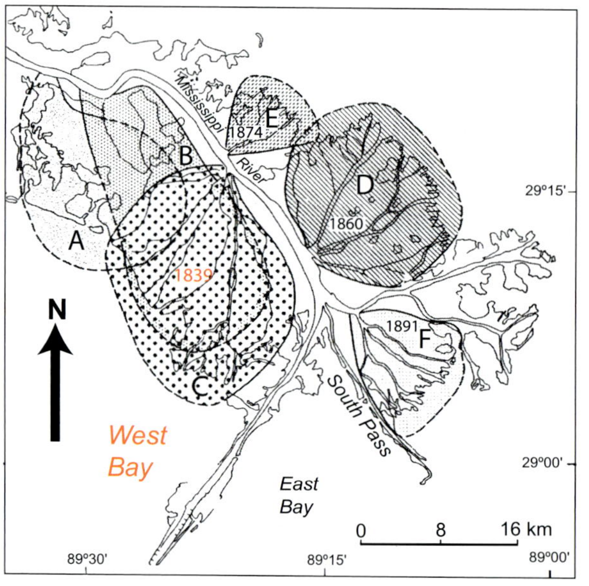

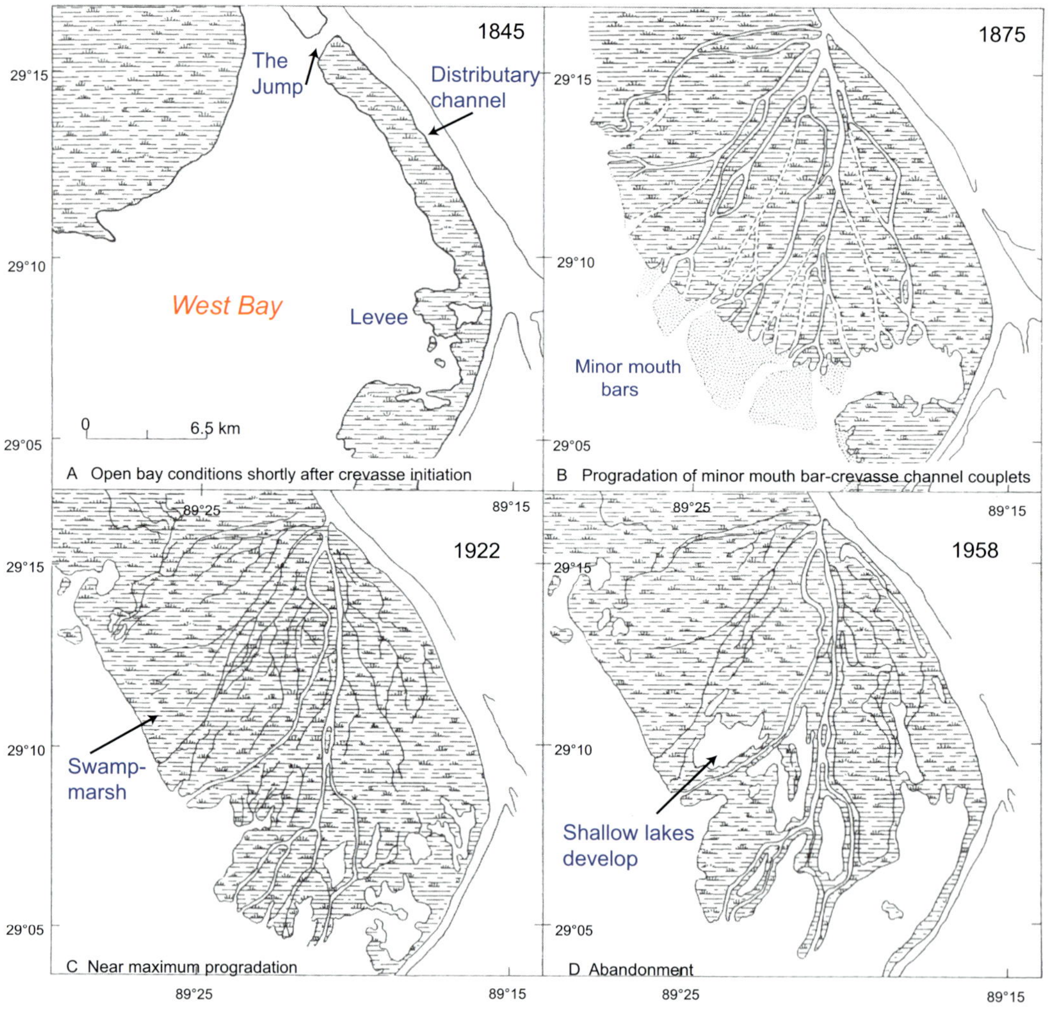

Figure 3.19 Recent evolution of the modern Mississippi delta showing periodic shifts of delta sub-lobes along the coastal margin. Upper panel: dates denote when bay-filling processes and sub-lobe formation of the lower delta began in the nineteenth century (after Coleman 1982, 1988). Lower panel: details of the cycle of initiation, progradation and abandonment of one sub-lobe that formed in West Bay (after Coleman 1982). Discussion in text.

mapped in the western bay, one flowing towards Heracleion and the other towards East Canopus before 2500 years before the present (Chen *et al.* 1992), while a third, younger channel was detected to the east in the central bay (El Bouseily and Frihy 1984; El Fattah and Frihy 1988). Traces of these three channels, now submerged and for the most part buried by sediment in the bay, have now also been recorded by the seismic profiles obtained in the present study (Figure 3.10). The gradual natural processes in Canopic sub-lobe evolution and continued morphological change, including the shift of low-lying channels and wetlands along the western bay coast, would have produced serious geohazards. In some cases, this would have involved displacement of large man-made structures and statuary to newer channel sites on younger, higher-elevation sub-lobes.

The selection of progressively younger channels receiving more active flow was essential, since the primary purpose of the Greek, Ptolemaic and Roman sites along the Canopic coast was for navigation and trade. Continuing changes in coastal marsh development and configuration would also have affected communication between the low-lying older and newer settlements. Travel in the sectors close to the cities would not have been by road, but rather by boat, in natural channels and artificially excavated canals. Considerable boat traffic can be envisioned between town centres that were located preferentially on topographically higher island-like lands scattered along the wetland-rich coastal margin bordered by a sandy shoreline. Although his description is vague, the historian Diodorus Siculus (*c.* 80–20 BC) suggested that pontoon boats may have been used in some channel mouth settings such as these.

Earlier, it was indicated that the two occupied sites were lowered from their original subaerial

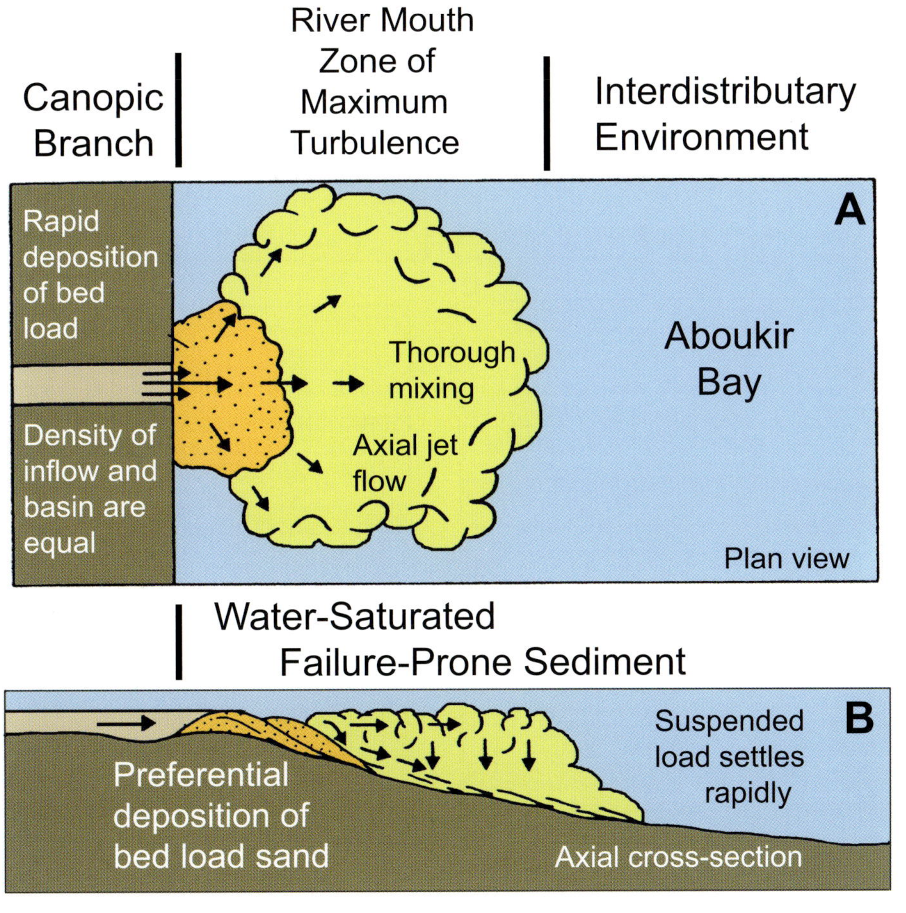

Figure 3.20 (A) Schematic showing hydraulic conditions, axial jet flow and sediment discharge at a river mouth under normal outflow conditions into a standing body of water such as Aboukir Bay. (B) Diagram (axial cross-section) of similar flow conditions, showing deposition of sand bed load at and just beyond river mouth, and rapid release and accumulation of finer-grained, water-saturated sediment somewhat farther basinward (modified after Elliott 1986 and others).

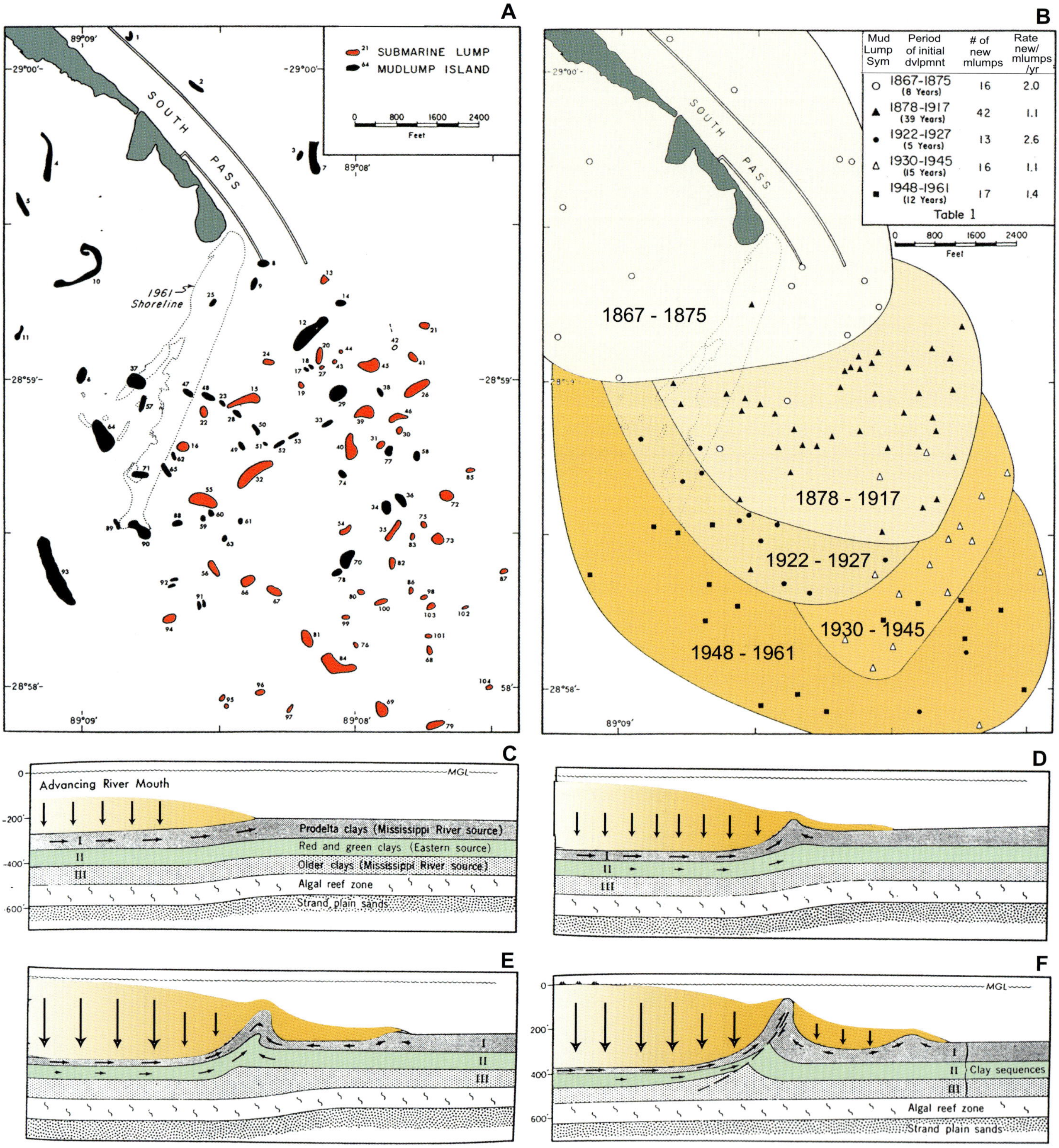

Figure 3.21 Formation of diapirs (also called mudlump islands) by the upward flow of sediment just off one of the mouths (South Pass) of the Mississippi (modified after Morgan *et al.* 1963). (A) Regional distribution of 104 submarine (in red) and subaerial (in black) diapirs at delta river mouth. (B) Sequence of Mississippi delta diapir formation from 1867 to 1961, or 104 diapirs in 94 years (more than one such feature per year), usually following floods; note seaward advance (to SSE) of diapir development. (C–F) Geological sections highlight syndepositional formation of diapirs by sediment overload and flowage of water-saturated deposits as the sand at the river mouth advances seaward. In origin, these compare with Canopic sub-lobe diapirs at Heracleion in Figure 3.9.

position at about +2 m above sea level to a present depth of ~6 m or more on the western bay-floor during the past 2500 years. The above-described gradual-process scenario (Figure 3.24, Category A), however, cannot fully account for the total elevation change of 8 m. A lowering of this magnitude, if near continuous during the past 2500 years, would have required a mean annual subsidence rate of at least 3.2 mm, or one much higher than is calculated on the basis of radiocarbon-dated core sections (1.7–2.0 mm per year). Moreover, a gradual lowering of the structures by 8 m at the calculated average rate of 2.0 mm per year would have required an additional 1500 years (at least 4000, not 2500, years). Thus, gradual natural processes, while significant overall, can account for little more than half of the total submergence (i.e. 4–5 m of the total 8 m in 2500 years). How, then, are we to explain that the ruins presently lie at a bay-floor depth of ~3.5 m deeper than would be expected if only eustatic sea level rise and gradual land lowering had been involved? It is apparent that other mechanisms have also contributed to the lowering of sites to their present depths.

7 Physical conditions leading to rapid substrate failure

In order to understand the subsidence of Heracleion and East Canopus and their additional submergence to a depth of 5 m or more in Aboukir Bay, it is critical to focus on attributes of sediment on which they were initially placed. It is particularly significant that both cities were positioned in close proximity to distributary mouths in the Canopic deltaic sub-lobes. As a consequence of this location, substrates immediately beneath the two centres were deposited as soft, water-saturated and underconsolidated sediment prone to failure and rapid subsidence. Moreover, diver excavation, coring and sub-bottom profiling do not record evidence of any large natural rock substrate or artificially constructed solid foundation, including kurkar, underlying the structures at the two centres.

A brief overview of sedimentological conditions at delta margins near river mouths follows. Active delta distributary channel mouths are particularly high-energy transitional points between land and sea. It has long been recognised (by, for example, Strabo in *c.* 7 BC, in his *Geography*) that landform configuration and depositional conditions at delta coasts are largely the consequence of interaction between river discharge and wave and other marine forces (tides, coastal currents) that prevail in the vicinity of a river mouth (Wright and Coleman 1973). Hydraulic characteristics in such environments are usually intense, even during normal river flow stage (Figure 3.20); conditions become more extreme at times of flood. Studies have compared river discharge patterns at delta mouths at low- and high-water stage to variable forms of jet flow (Bates 1953). The delta mouth settings, especially during high flow and extreme high flood stages, are characterised by markedly altered hydraulic changes that involve much increased outflow velocities, bed shear and fluid turbulence along with increased, and commonly prolonged, discharge of denser sediment-laden river waters (Wright and Coleman 1974).

Turbulence and sedimentation rates are highest at and just seaward of a delta river mouth (Figure 3.20). Current velocities diminish seaward and along the margins of the jet flow, resulting in release and deposition of sandy bed load seaward of the river mouth, and of water-enriched suspended sediments somewhat further from the fluvial outlet. Such rapidly deposited materials at the delta front are usually characterised by excess of pore water pressure between particles, especially clay and fine silt. This pressure results in low sediment strength and underconsolidation. Moreover, entrapment of organic matter along with fine-grained particles released at river mouths commonly leads to the formation of methane gas, which also contributes to a generation of excess pore pressure within the delta front deposits. As a result of these conditions, the rapidly deposited sediments that accumulate seaward of delta mouths are especially prone to failure (Morgan *et al.* 1963; Coleman 1982, 1988; Maestro *et al.* 2002).

Our geological analyses of delta mouths in Aboukir Bay record the three-dimensional configuration of sediment that originally had large excess pore water pressure and appears to have accumulated in aerially restricted areas. They occupy an area where deposits accumulated most rapidly, and where cumulative loading (weighting) effects on underconsolidated materials would have prevailed. For example, the addition of a new layer of sediment added during a flood can induce a change in the particle-to-particle configuration of the underlying deposit, and this often leads to sediment flowage (Figure 3.21C–F) and dewatering of the underlying clays and silts (Morgan *et al.* 1963). Rapid or sudden failure, for example, can occur where water-saturated muds fail and are squeezed upward by differential loading (Figure 3.21D–F), especially by the addition of new sandy deposits such as those concentrated along the advancing river channel and seaward of the mouth (Coleman and Wright 1975; Galloway 1975; Coleman 1982, 1988).

Soil engineers and geologists have determined that excess pore pressure can develop in a stratum

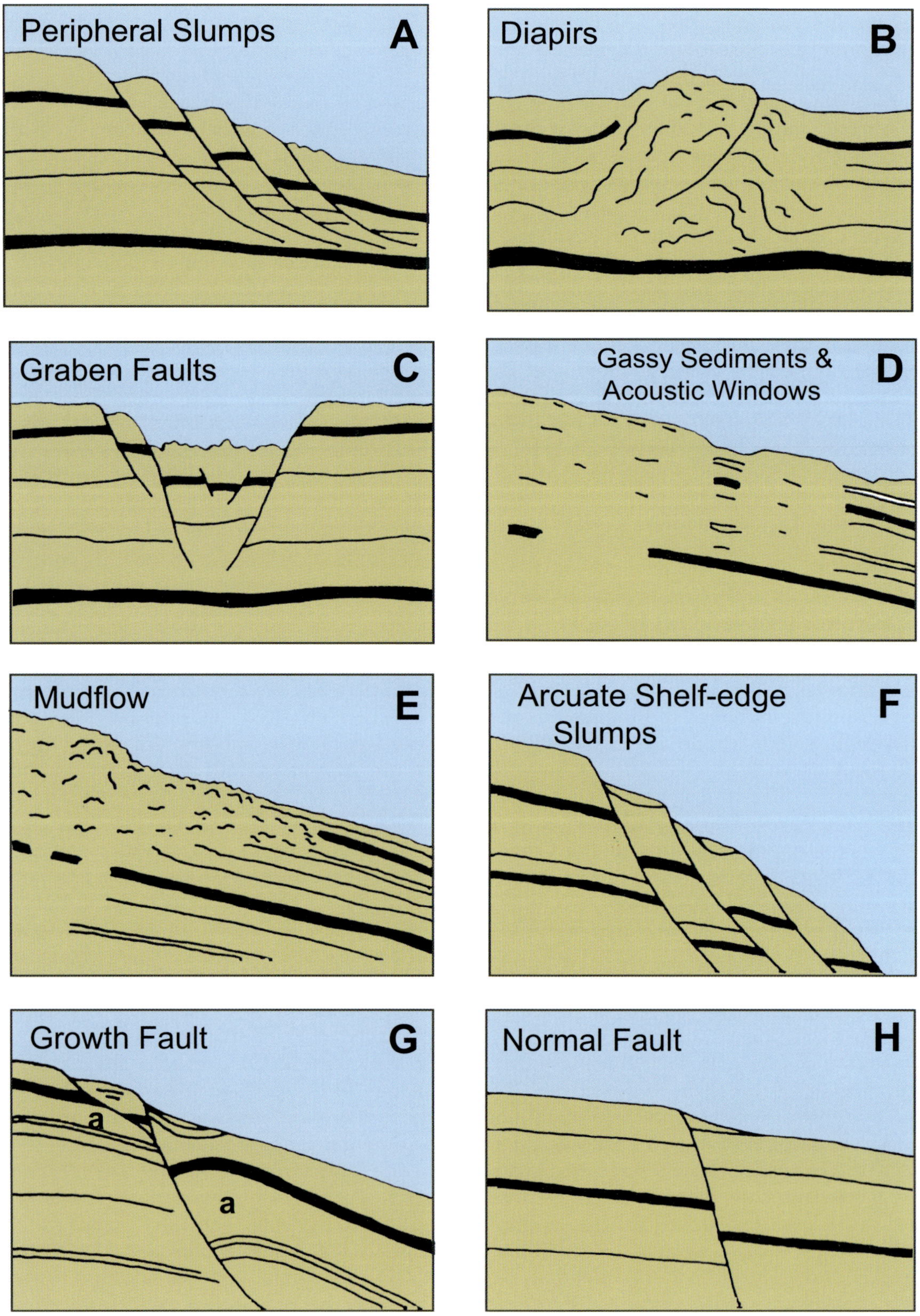

Figure 3.22 Examples of diverse sediment failure and mass movement commonly recorded at different delta margins (after Coleman 1982). Some of these phenomena resulted in failure and lowering of Canopic sub-lobe margin sediment by at least 2–3 m or more to the basin floor just seaward of the coast.

that is increasing in thickness owing to the rapid addition (weighting) of overlying deposits. As a result of such build-up, this stratum can fail suddenly by slumping (Figure 3.22A,F) on minimal slopes (<1–2 degrees), and even on a horizontal surface (Figure 3.23). Sea-floor topographic surveys in front of delta mouths, including those in tectonically tranquil regions, commonly reveal a 'stairstep' profile (Figure 3.23) that results from rotational slumping, mud flows (Wright and Coleman 1974; Coleman 1982; Figure 3.22E), and growth and normal faults (Maestro *et al.* 2002; Figures 3.22G,H, 3.8, 3.23). Studies in various delta regions have shown that these phenomena commonly cause sudden displacement of sediment masses from both the subaerial and shallow submerged natural levees near outer delta margins (elevations of +2 to -2 m) to the adjacent sea-floor surface. This can account for basinward motion and a lowering of sediment by several metres or more (Figures 3.22, 3.23B).

Changes of sediment physical properties that lead to sudden failure at a delta mouth (Figures 3.21–3.23) can thus explain how poorly consolidated deposits could subside rapidly seaward of the river mouth, and why the ruins in the study area are now observed at depths of -5 to -7 m at the two localities. It is also postulated that differential loading by distributary mouth sands near the river mouths not only caused flowage and triggered failure of sediment, but also caused tilting (Figure 3.9) and physical disruption of underlying Holocene strata (Figures 3.18D,E). In fact, the seismic sub-bottom profiles we have examined

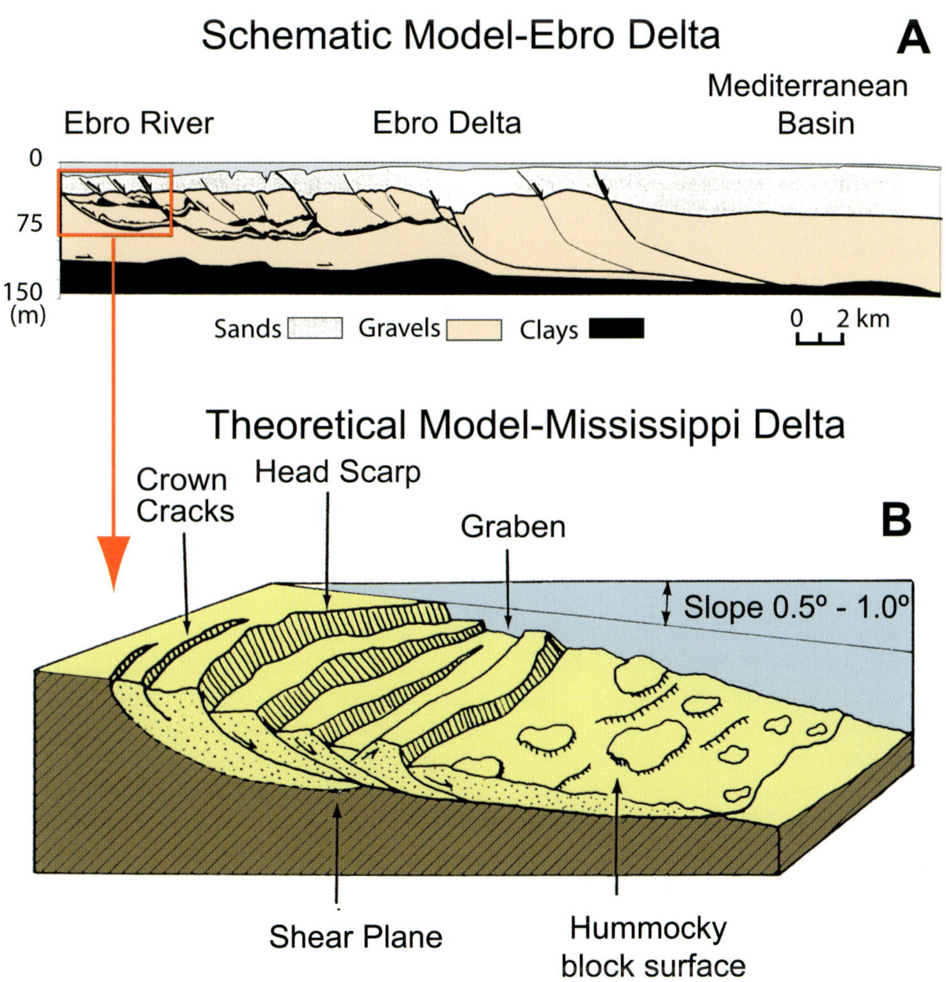

Figure 3.23 (A) Growth fault offset and massive syndepositional failure and lateral displacement of sediment (see arrows) at and seaward of the modern Ebro delta in the western Mediterranean (modified after Maestro *et al.* 2002). (B) Growth faulting and associated sediment failure at the Mississippi delta margin results in a steplike profile from land to basin; compare crown crack features here, for example, with trench discovered at East Canopus (Figures 3.12, 3.13).

show the probable deformation of sediment sections into diapir domes (compare Figures 3.9D and 3.21D–F). This assemblage of syndepositional features does not occur randomly in Aboukir Bay, but is concentrated near and just seaward of channel mouths. As in the case of the Mississippi, diapiric dome features and associated tilted beds of Canopic sub-lobes appear to have formed rapidly during outgrowth of channels (Figure 3.21A,B), and also by loading and sediment flowage associated with the seaward build-out of the delta margin (cf. Morgan *et al.* 1963). As a consequence, disturbed and fluidised strata extend to the sea-floor surface near the former mouths of the Canopic branch (compare Figures 3.9, 3.11 upper panel, and 3.18D,E with Figures 3.21C–F and 3.23B).

Moreover, it is quite probable that even more localised effects of subsidence at Heracleion and East Canopus resulted from emplacement of heavy structures, such as temple columns and large walls, directly on the soft mud substrate. In many modern coastal cities (New Orleans, Venice, Shanghai, Bangkok and others), there are numerous examples of rapid settling of homes and buildings where such structures have been set on water-saturated clay and silt. Even when positioned on special raft foundations, buildings underlain by soft mud have rapidly subsided by 3 m or more in less than a century (Waltham 2002). Recall that there is no evidence of pilings or foundations having been placed beneath dense granite columns, walls and other large structures at the two cities in the study area. As a result, some almost certainly would have rapidly settled and/or tilted under their own weight, requiring periodic readjustment of such features, or their abandonment, even before the overall settlement area finally subsided beneath the waves.

In summary, we can relate features actually observed and concentrated at and near the former sites and their adjoining Canopic channel mouths in Aboukir Bay with phenomena of sediment failure and syndeposition discussed in this section.

- Growth and normal faults, crown cracks, and large syndepositional features at and near the sea-floor (Figures 3.21–3.23) are denoted by distinct magnetic anomalies at both sites (Figure 3.12), diapirs (Figures 3.7, 3.9), and the buried trench at East Canopus (Figures 3.12B, 3.13).
- Exposure of strata of mid-Holocene or older age (diapir feature) occurs locally on the modern bay-floor surface in the vicinity of the sites, as noted on side-scan images (Figure 3.11) and on seismic profiles (Figures 3.7, 3.8, 3.9B,D,E).
- Larger proportions of sand (probable channel mouth deposits) are recorded in cores recovered at these localities (Figure 3.17).
- Contorted sediments with typical fluidised structures in cores are recovered primarily at and near the two submerged cities (Figures 3.7, 3.8, 3.17, 3.18D,E).
- Radiocarbon dates from the top of cores and diver-recovered surficial samples commonly record a surprisingly old age for near-surface sediment (to >5000 years before present) rather than expected late Holocene dates of <3000 years before present (Figure 3.17).
- Some upper sediment samples of Holocene age comprise fauna that are primarily of brackish and freshwater origin, rather than of expected open marine biofacies (palaeontological analyses in progress).

It is the overall assemblage of the features listed above that best provides a means to comprehend causes and effects of the more rapid to sudden changes at the sea-floor that, in turn, appear to have affected adversely the two settlements, including their damage and submergence.

8 Triggering of sudden failure and subsidence at Canopic mouths

The scenario we envision for submergence of Heracleion and East Canopus involved both the gradual relative sea level rise in Aboukir Bay and the effects of rapid accumulation of water-saturated sediment and episodic sediment failure at Canopic mouths. Together, these processes account for most of the subsidence of the ruins to their present depth (Figure 3.24, Categories A and B). The results of geological exploration, together with analysis of the archaeological data, indicate that some destruction of the settlements and their subsidence occurred

Facing page: **Figure 3.24** Summary scheme that highlights the causes and effects of both (A) gradual and (B) rapid to sudden (including catastrophic) events affecting western Aboukir Bay during the past ~2500 years. Emphasised are several natural processes that led to submergence of the ruins of Heracleion and East Canopus to depths of 5–7 m below sea level, from ~500 BC to the present. Discussion in text.

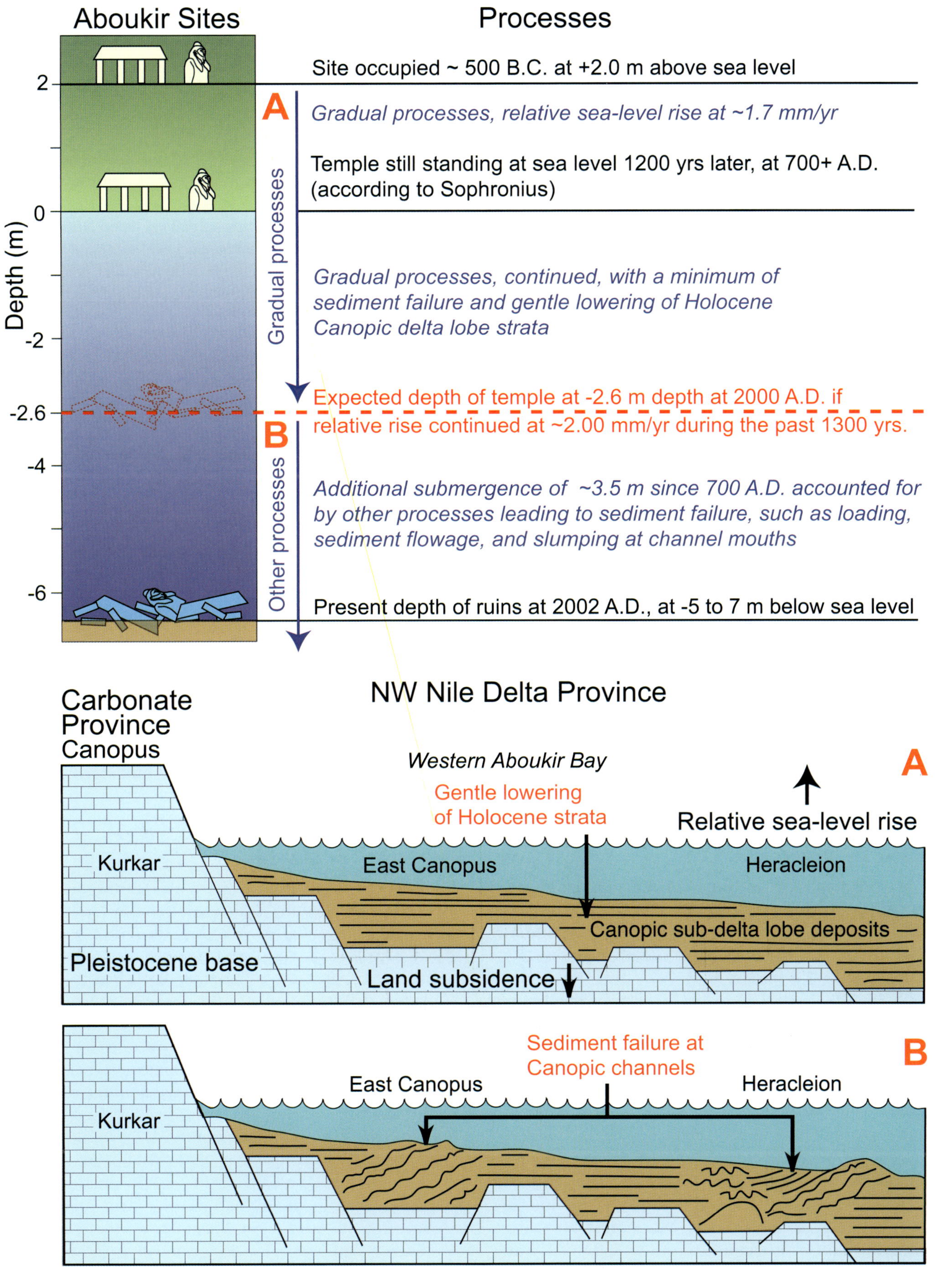
Aboukir Sites
Processes
Depth (m)
2
0
-2
-2.6
-4
-6
Site occupied ~ 500 B.C. at +2.0 m above sea level
A
Gradual processes, relative sea-level rise at ~1.7 mm/yr
Temple still standing at sea level 1200 yrs later, at 700+ A.D. (according to Sophronius)
Gradual processes
Gradual processes, continued, with a minimum of sediment failure and gentle lowering of Holocene Canopic delta lobe strata
Expected depth of temple at -2.6 m depth at 2000 A.D. if relative rise continued at ~2.00 mm/yr during the past 1300 yrs.
B
Other processes
Additional submergence of ~3.5 m since 700 A.D. accounted for by other processes leading to sediment failure, such as loading, sediment flowage, and slumping at channel mouths
Present depth of ruins at 2002 A.D., at -5 to 7 m below sea level
Carbonate Province Canopus
NW Nile Delta Province
Western Aboukir Bay
A
Gentle lowering of Holocene strata
Relative sea-level rise
Kurkar
East Canopus
Heracleion
Canopic sub-delta lobe deposits
Pleistocene base
Land subsidence
Sediment failure at Canopic channels
B
East Canopus
Heracleion
Kurkar

abruptly. Strong evidence is provided by human skeletal remains beneath damaged structures (such as walls). Moreover, populations almost certainly would not have left gold, gems, statuary and other valuables in place (cf. Goddio F, this series) if gradual sea level change and land lowering alone had been raising the water level on to the site at a long-term mean rate of about 2.0 mm per year.

The new geological findings presented here indicate episodes of sudden subsidence that resulted from prevailing conditions of rapid sediment accumulation, depositional loading and oversteepening. It was the periodic failure of unstable sediment masses at Canopic river mouths that resulted in their displacement to depths of 3 m or more in the deeper adjacent Aboukir Bay floor. Findings such as magnetometer anomalies and the associated artificially buried trench at East Canopus discovered by IEASM archaeologists (Figures 3.12B, 3.13) can now be interpreted as sea-floor surface expressions of disturbed sediment masses, crown cracks and/or growth faults of the type formed by rapid to sudden failure (cf. Coleman 1982; Figures 3.21–3.23). From our findings to date, it appears that physical conditions associated with rapid sedimentation, including water saturation and high pore pressure, loading and oversteepening, could have produced substrate failure, even in areas not affected by earthquake tremors or sudden loading by tsunamis and storm surges.

In addition to the periodic failure of oversteepened and water-saturated sections of unstable deposits, what other natural phenomena or pulses could have triggered sediment failure? It is reasonable to expect that these river mouth settings were influenced primarily by the annual floods of the river Nile. These floods, so critical for the development of Egyptian civilisation (Butzer 1976; Hassan 1981), occurred each year at about the same time, in late summer and autumn, and their levels were carefully recorded (Popper 1951). Some of the annual floods were exceptionally high as a result of palaeoclimatic fluctuations in central and eastern Africa (Hurst 1931–66; Ball 1939; Bell 1970; Riehl and Meitin 1979; Hassan 1981; Hamid 1984; Shahin 1985; Said 1993). Such floods caused sudden additional loading on already unstable deposits, and consequently were the most likely and frequent triggers of sediment failure at and near the low-lying site localities. High flood events drowned large parts of the Canopic delta margin and, in some instances, induced a sudden switch of delta lobe channels. It is proposed that here, as at the margins of some other of the world's large deltas, flooding was a major factor that caused sediment flowage and formation of diapirs (Figures 3.9D, 3.21) and fluidised structures (Figures 3.17, 3.18E).

More specifically, it has recently been suggested that high Nile floods in the mid-eighth century AD triggered sediment failure that caused the demise of East Canopus (Stanley *et al.* 2001, 2004). Among the factors used in support of this hypothesis is the recovery at these ruins of gold coins, of which the two most recent date from Arabic times, 729–730 and 718–719 (Goddio F, 2007). A particularly high Nile flood (at least 1 m higher than normal high floods) in 741 or 742 AD was recorded at the Roda Nilometer in Cairo (Popper 1951). Other floods of this magnitude are only recorded well before (719 or 720 AD) and long after (805 AD) the 741/742 AD event. It is probable that some major destruction and lowering of East Canopus is related more closely to this documented high flood, rather than to any other undocumented event (earthquake, tsunami, storm surges) 12> centuries ago (Stanley *et al.* 2001, 2004). In earlier time, such high floods would probably also have affected Heracleion, causing periodic channel migration, substrate failure and serious damage to structures.

The importance of the high Nile floods and their effects on the two centres in the bay does not exclude the role of earthquake tremors (even low-magnitude events) and effects of sudden loading by tsunamis and storm surges. These events, in some instances, could have served as triggers of sediment failure. There has been a concerted effort to document important ancient earthquake and seismicity-related phenomena in the eastern Mediterranean (Guidoboni 1994; Soloviev *et al.* 2000). Such records indicate that earthquakes and tsunami events were important factors in the destruction of Alexandria (Stanley and Bernasconi 2006), a somewhat seismically more active region to the west of Aboukir Bay; earthquake damage has been proposed for structures built in Alexandria's eastern harbour as well as for part of the city along the harbour coast (Guidoboni 1994). It is conceivable that at times micro-earthquakes and occasional powerful tremors also affected the Aboukir Bay study area but, as yet, no firm correlation has been established between seismicity, earthquakes and subsidence. Moreover, it is unlikely that a natural devastating event such as an earthquake or a tsunami which could have destroyed these cities would have gone unrecorded in historic accounts.

As an example, we note the wave damage related to an earthquake that occurred on the Levantine margin in 746 BC (possibly 743 or 745 BC; Guidoboni 1994; Soloviev *et al.* 2000), but no earthquake activity is recorded in Egypt during this period. Geological investigations and seismo-archaeological surveys in sectors of the eastern Mediterranean have also suggested that the period from the fourth to the sixth century BC was one of unusual clustering of destructive earthquakes (Pirazzoli 1986; Stiros 2001). It is possible that tremors, and tsunamis associated with these events, played a role in modifying the configuration of the delta on the north-western margin of Egypt (Stanley and Bernasconi 2006).

It would be unreasonable, however, to overlook the fact that the annual flood of the Nile was a regular and powerful pulse that could have induced failure, particularly at the low-lying river mouths where the two settlements were positioned. It is critical to recognise that the Holocene sediment in the bay is disturbed, fluidised, tilted and offset specifically beneath the ruins of the ancient cities and the adjacent channels and river mouths (Figure 3.9D). This disturbed sediment, so obviously visible in the seismic sub-bottom and side-scan sonar data at and near Heracleion and East Canopus, is not apparent in the surrounding region away from the sites. For this reason, we believe that these syndepositional features would most commonly have been associated with flood events. We expect, on the other hand, that the effects of earthquakes and tsunamis should be observed not only in the Canopic deposits forming the substrate at the river mouths and beneath the ruins of the two ancient cities (Figures 3.9D, 3.24B) but over a much broader area of Aboukir Bay (cf. Wheeler 2002). Moreover, our examination of the contorted strata indicates that they more closely resemble fluidised flood deposits rather than strata laid down by tsunamis (Fujiwara *et al*. 2000; Takashimizu and Masuda 2000). Gradual relative sea level rise that led to shifts of delta sub-lobes, combined with the effects of sudden sediment failure in this region, almost certainly caused periodic displacement of settlements at the Canopic delta margin.

The physical processes may have occurred as episodic events instead of a single catastrophic failure. These events probably resulted in lowering of some settlement sectors below the waves in the bay, while other parts of the cities remained inhabited on low islands above water, perhaps as described by Herodotus. The recovery of archaeological artefacts such as Byzantine and Arabic coins at both Heracleion and East Canopus indicates that habitation of some of these islands may have continued for decades, or even centuries, after submergence and abandonment of major sectors of the original cities. The periodic shifts of occupied locales suggest that there may be ruins yet to be discovered in Aboukir Bay.

As our geoarchaeological investigations continue, future discoveries will shed more light on the origin and timing of triggering events that resulted in the failure of the sediment beneath the two ancient centres and their associated cultural impacts. Most Greek and Roman buildings in the circum-Mediterranean region impress us by their beauty and design. They were obviously built to last. It is thus surprising to discover some similar structures now submerged in the western part of Aboukir Bay. It seems difficult to accept that the master architects had placed such large, heavy structures directly on soft, unconsolidated and unstable sediment. Surely, the Egyptian, Greek and Roman builders and designers would have been aware of such engineering and constructional problems and associated geohazards. By using the available, although limited and restricted, space for their city placement, they in effect created 'an accident waiting to happen'. If the Egyptians, the Greeks and subsequent inhabitants initially built on the Aboukir Bay margin it was for defensive and economic purposes, including tolls and associated beneficial trade payments along the Canopic branch of the Nile. Given their traditional long-lasting architectural attributes, it is apparent that in this particular case a much lower priority was placed on building site location and substrate characteristics.

Chapter 4

Faunal Analyses in the Interpretation of the Submergence of Substrates beneath Heracleion and East Canopus

Maria Pia Bernasconi, Romana Melis,
Nevio Pugliese, Jean-Daniel Stanley and Alessio Bandelli

'I have seen that Egypt projects into the sea beyond the neighbouring land, and that seashells show up on mountains, and that brine-salt comes to the surface.' Herodotus, *The Histories*, 2.12

1 Introduction

Information on submergence of the cities in Aboukir Bay and failure of their sediment substrate has been obtained as a result of recent geological and geophysical surveys. These provide new insight on the triggering mechanisms as well as how and when the two cities subsided in Aboukir Bay (Stanley *et al.* 2001, 2004; Chapter 3 in this volume).

The purpose of this study is twofold: (1) to determine if correlation using cores can be established between now submerged substrates in different geographical localities at and near the submerged sites; and (2) to interpret the environmental history recorded in cores to gain additional information on the nature of lowering of the sediment substrates upon which the two ancient cities were originally built. An investigation focusing on these two aspects will probably provide some clarification of sea level changes and syn- and post-depositional conditions affecting these substrates during the past 2500 years. To achieve these goals, we examined various fauna from a series of radiocarbon-dated vibracores collected at nine localities beneath the two cities (Figure 4.1).

Emphasis herein is on molluscan assemblages, including bivalves, gastropods and scaphopods, and on microfossils, including foraminiferal and ostracod fauna. The approach used is based on the *a priori* assumption that most fauna encountered in Holocene core sections recovered in the study area presently live in the Mediterranean. Thus, an assessment of molluscan and microfossil assemblages in core sections should shed new light on depositional conditions through time on the mobile Nile delta margin. The investigation focuses on faunal changes through time, i.e. those occurring prior to, during and following subsidence of both cities into the sector that is now a fully marine bay.

2 Methodology

2.1 General

The fauna were examined in 9 of the 17 vibracores recovered in Aboukir Bay in 2001 (Jorstad and Stanley, Chapter 5 in this volume). These mid-to-late Holocene sections were selected on the basis of representative lithofacies in the borings and on their geographic location relative to the two submerged sites (Figure 4.1). The selected cores at and near Heracleion are ST1, ST2, ST3, ST7 and ST9, and those at and near East Canopus are ST12, ST13, ST14 and ST17. Core sections range in length from 321 to 549 cm, and the nine sampled cores average 430 cm. The cores were split into two halves, described, X-radiographed and sampled at the National Museum of Natural History (NMNH) at the Smithsonian Institution (Washington, DC).

Detailed logs of all Aboukir Bay cores are depicted in Jorstad and Stanley (Chapter 5, this volume), and simplified lithologic logs of the nine cores examined in this study are shown in Figures 4.2 and 4.3. Shown on these logs are 26 radiocarbon dates given in uncalibrated years before present (yrs BP). Most of the radiocarbon dates were obtained by AMS analyses (Beta Analytic, Inc., Miami, Florida) using plant material (database listing in Jorstad and Stanley, their Table 5.1, Chapter 5 in this volume).

Bulk samples (~100 cm^3, corresponding to core sections about 2 cm thick) from the nine cores were examined for grain size (79 samples), molluscan fauna (73 samples) and microfossils (78 samples). The number of samples collected per core ranges from 6 to 14, and their depths are shown on the simplified logs of the nine cores (Figures 4.2, 4.3). Texture of each sample was determined using a Coulter Counter LS200 particle analyser. The grain size database for each sample includes: percent

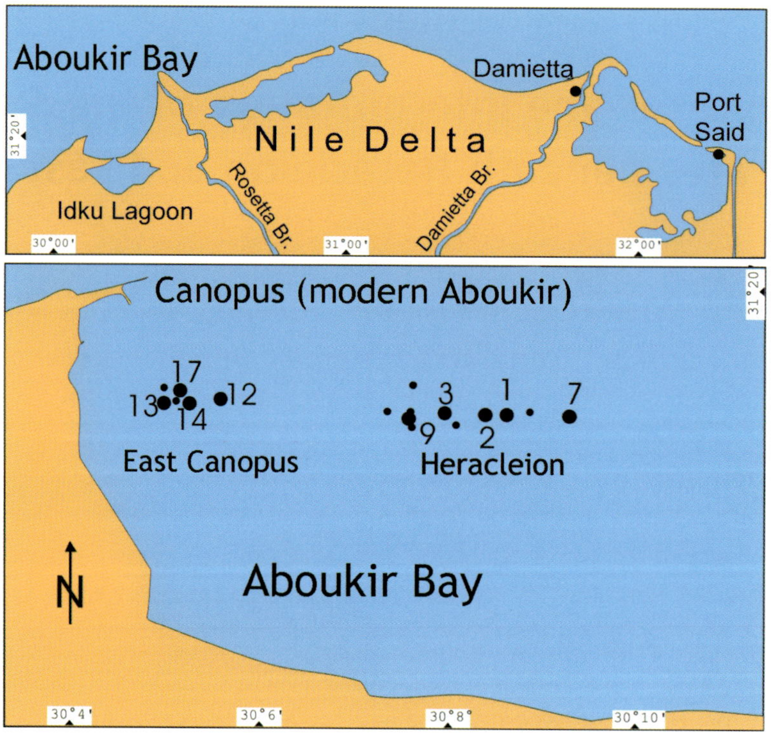

Figure 4.1 (Top) Lower Nile delta margin of Egypt. (Bottom) Map of western Aboukir Bay showing locations of vibracore sites at Heracleion and East Canopus.

Table 4.1 Textural data for samples examined in nine cores from Heracleion and East Canopus

ST 1	*Depth (CM)*	*Sand%*	*Silt%*	*Clay%*	*Mean (ÌM)*	*Median (ÌM)*	*Mode (ÌM)*	*Skewness*	*Kurtosis*
1	38 - 40	0.75	88.25	11.00	17.65	13.54	34.58	0.600	-0.054
2	103 - 105	1.19	86.64	12.17	17.40	12.40	28.69	1.150	0.058
3	163 - 165	5.66	80.57	13.77	22.31	10.32	16.40	3.730	14.800
4	223 - 225	10.53	77.03	12.44	32.08	13.75	26.14	3.050	9.130
5	273 - 275	19.97	70.06	9.97	43.65	18.79	31.50	2.080	3.680
6	333 - 335	39.70	52.90	7.40	95.66	31.19	223.40	0.664	-0.274
7	433 - 435	91.67	7.50	0.83	200.20	200.00	203.50	-0.047	0.589
8	453 - 455	59.61	38.15	2.24	143.70	89.67	203.50	3.420	16.900
9	483 - 485	39.03	55.85	5.12	88.84	38.90	116.30	3.820	21.000
10	533 - 535	26.27	65.69	8.04	67.11	25.59	34.58	4.940	34.900
ST 2									
1	33 - 35	99.20	0.71	0.09	588.10	445.20	324.30	1.330	1.220
2	93 - 95	99.04	0.83	0.13	535.10	426.60	356.10	0.000	2.000
3	143 - 145	100.00	0.00	0.00	346.10	287.40	269.20	3.140	12.100
4	223 - 225	99.99	0.01	0.00	363.90	294.80	269.20	2.790	9.230
5	310 - 312	98.29	1.48	0.23	337.70	261.80	245.20	2.980	10.400
ST 3									
1	23 -25	99.06	0.85	0.09	531.80	354.90	269.20	1.450	1.400
2	53 - 55	96.60	2.80	0.60	488.10	340.50	269.20	1.530	1.960
3	103 - 105	80.08	17.14	2.78	297.90	244.80	295.50	2.270	6.940
4	148 - 150	11.94	79.15	8.91	31.20	24.31	41.67	1.780	4.690
5	183 - 185	12.60	77.65	9.75	30.29	23.30	41.67	1.300	1.900
6	228 - 230	2.76	86.01	11.23	19.91	13.46	31.50	1.040	0.243
7	283 - 285	2.85	86.97	10.18	22.31	18.03	34.58	0.805	-0.134
8	343 - 345	0.94	87.67	11.39	18.32	14.51	28.69	0.965	0.384

Table 4.1 Textural data for samples examined in nine cores from Heracleion and East Canopus (continued)

ST 7	*Depth (CM)*	*Sand%*	*Silt%*	*Clay%*	*Mean (ÌM)*	*Median (ÌM)*	*Mode (ÌM)*	*Skewness*	*Kurtosis*
1	70 - 72	98.91	0.99	0.10	277.90	239.60	245.20	3.700	18.000
2	103 - 105	25.45	62.62	11.93	93.37	14.94	223.40	3.130	12.500
3	140 - 142	49.84	41.69	8.47	133.60	61.33	245.20	0.628	-0.815
4	223 - 225	96.48	2.51	1.01	266.70	235.70	223.40	3.260	15.500
5	288 - 290	98.06	1.50	0.44	316.20	261.00	245.20	3.250	13.800
6	343 - 345	98.89	1.00	0.11	324.90	269.90	245.20	2.330	7.020
7	403 - 405	99.06	0.83	0.11	284.80	262.40	269.20	2.850	15.800
8	463 - 465	100.00	0.00	0.00	274.60	243.90	245.20	3.960	20.900
9	543 - 545	99.46	0.42	0.12	254.20	226.10	223.40	3.870	21.800
ST 9									
1	13 -17	92.76	6.57	0.67	694.30	547.70	1091.00	0.683	-0.577
2	27 -29	23.17	66.81	10.02	108.30	17.50	18.00	3.050	8.930
3	90 - 92	84.74	13.88	1.38	152.70	108.30	105.90	5.240	32.200
4	123 -125	2.73	84.77	12.50	18.81	13.86	26.14	1.680	4.070
5	193 -195	69.90	27.85	2.25	168.40	91.24	87.90	3.850	16.400
6	273 - 275	92.47	6.64	0.89	158.70	131.80	127.60	4.540	27.600
7	363 - 365	50.55	41.90	7.55	128.50	65.94	185.30	3.820	20.100
8	383 - 385	90.68	8.22	1.10	722.50	613.50	14443.00	0.491	-0.998
ST12									
1	33 - 35	0.00	88.94	11.06	13.77	10.40	26.14	0.846	-0.267
2	123 - 125	0.50	87.02	12.48	16.65	12.68	28.69	1.020	0.546
3	223 - 225	0.62	86.30	13.08	15.31	11.03	26.14	4.200	34.500
4	373 - 375	13.13	76.98	9.89	45.32	17.72	31.50	4.030	21.000
5	423 - 425	14.02	71.84	14.14	47.71	8.92	9.37	2.520	5.610
ST 13									
1	15 -16	98.84	1.03	0.13	559.90	390.60	269.20	1.250	0.916
2	35 - 37	40.53	53.62	5.85	109.40	44.20	50.23	3.030	12.200
3	54 - 55	6.97	78.44	14.59	26.56	9.51	14.94	3.450	11.000
4	85 - 87	16.42	75.89	7.69	50.49	15.89	18.00	3.320	14.300
5	107 - 109	7.21	82.86	9.93	24.32	14.67	28.69	2.710	8.820
7	131 - 133	9.55	80.68	9.77	36.72	12.71	16.40	3.780	16.500
8	138 - 140	40.54	54.79	4.67	182.60	34.79	269.20	1.970	3.730
9	164 - 166	98.41	1.44	0.15	695.00	483.20	223.40	0.743	-0.708
10	192 - 194	13.40	77.64	8.96	26.97	14.65	16.40	1.780	2.840
11	232 - 234	0.53	86.12	13.35	12.97	9.16	16.40	1.630	3.530
12	264 - 266	0.00	84.07	15.93	9.43	6.98	10.29	1.000	0.173
13	293 - 295	0.00	84.46	15.54	10.56	8.14	13.61	0.887	-0.077
14	318 - 320	0.15	90.23	9.62	15.09	11.88	18.00	1.150	1.040
15	348 - 350	18.58	64.74	16.68	63.73	8.30	7.78	3.120	14.800
ST14									
1	18 - 20	97.62	2.14	0.24	405.20	297.60	269.20	2.390	5.840
2	103 - 105	5.19	84.92	9.89	23.92	14.92	28.69	4.230	23.200
3	203 - 205	0.42	85.60	13.98	14.70	9.72	28.69	3.830	33.100
4	293 - 295	7.73	81.03	11.24	27.56	13.17	26.14	3.260	10.800
5	363 - 365	0.00	86.40	13.60	11.66	8.34	13.61	1.260	1.180
6	443 - 445	47.39	44.19	8.42	114.70	32.53	223.40	0.575	-0.938
ST 17									
1	13 - 15	97.84	1.96	0.20	411.60	295.90	269.20	2.220	4.950
2	58 - 60	80.55	16.34	3.11	646.80	534.80	993.50	0.615	-0.554
3	88 - 90	95.21	4.14	0.65	766.40	690.30	993.50	0.479	-0.710
4	113 - 115	92.24	6.90	0.86	761.40	797.10	993.50	0.252	-0.958
5	161 - 163	4.42	84.29	11.29	18.98	11.12	23.81	2.990	11.100
6	252 - 254	2.79	86.83	10.38	19.54	13.80	28.69	1.500	3.070
7	343 - 345	0.00	83.98	16.02	12.00	8.91	26.14	0.877	-0.187

sand (>63 μm), silt (0.4–63 μm) and clay (<0.4 μm); mean, median and modal size (in μm); standard deviation, skewness and kurtosis (Table 4.1).

Molluscs, foraminifers and ostracods were separated from the terrigenous fraction using a 0.5 mm sieve for molluscs, and a 0.063 mm sieve for microfauna. The fauna were identified, statistically analysed and interpreted independently by the authors of this chapter: molluscs by M.P. Bernasconi, foraminifers by R. Melis, and ostracods by N. Pugliese and A. Bandelli. Statistical analysis was performed using the software PRIMER v5 (Plymouth Marine Lab.; Clarke and Warwich 2001). For this analysis, abundance data pertaining to molluscs, foraminifers and ostracods have been double-squared root transformed, and used for hierarchical agglomerative clustering based on Bray–Curtis similarity measure.

In the present study, the term lagoon is used broadly to designate not only lagoons in the usual sense, but also the associated wetland environments including marsh, tidal flats and inlets/outlets to the sea.

The syntheses and conclusions presented here were prepared by the five chapter authors.

2.2 Molluscs

Species were identified, and their abundance (number of specimens of each taxon in a sample) was determined following the method of Di Geronimo and Robba (1976), and also reported in Stanley and Bernasconi (1998). The 117 molluscan species identified are listed in systematic order in Table 4.2. Selected species were photographed by a scanning electron microscope (Steroscan 360 Cambridge Instrument) at the University of Calabria. All species identified in this investigation presently live in the Mediterranean, and in most cases the environments in which they occur are known. Studies pertaining to molluscs on Egypt's Nile delta margin, in modern settings as well as in Holocene sequences, include those of Pallary (1911), Bernasconi *et al.* (1991), Bernasconi and Stanley (1994, 1997) and Stanley and Bernasconi (1998). The bionomical/ecological information of taxa, defining the attribution of specific taxon to specific environment, is simplified (Table 4.3) from the letter-coded classification originally proposed by Pérès and Picard (1964). Herein, we assign three letter codes, and these are related to the equivalent ones defined by Pérès and Picard as shown below:

- marine (M) = AP, SFBC, C, AP-HP, DC, PE, SFHN , SVMC, HP-DC, HP, SGCF
- lagoonal (L) = LEE
- freshwater (FW).

A database was compiled (Table 4.3) using only those samples where molluscan taxa could be identified to species level, and only those samples that contain more than one species. In this matrix, the abundance value for each taxon is indicated. This shortened table contains 98 species in 33 samples and provides a matrix for multivariate analysis (cluster analysis: Bernasconi and Stanley 1997; Basso and Corselli 2002). The cluster plot, shown in Figure 4.4A, enables us to select natural groups of molluscan-bearing samples that define natural marine (M), lagoonal (L) and freshwater (FW) environmental settings.

In addition to the above 46 samples with identifiable molluscs, 27 (37%) of the 73 samples either (1) did not contain any molluscan fauna (denoted by the letter n = no molluscs), or (2) comprised only unidentifiable fragments (recorded by frag in Figures 4.2 and 4.3), or (3) in the case of two samples (ST13–12 and ST13–13) provided no sand-size residue of either organic or inorganic fractions.

2.3 Foraminifers

Where foraminifers are abundant, samples were subdivided using a dry splitter until an aliquot containing approximately 300 specimens was obtained. In the case of very low density of specimens (usually in sandy sediment), a flotation method using carbon tetrachloride (CCl4) was used to concentrate the tests. After an initial qualitative analysis, species counts were performed and recorded as number of specimens of each taxon. Selected specimens were examined and photographed by a Leica Scanning Electron Microscope at the University of Trieste. Identification of foraminifer species follows the taxonomy proposed by Le Calvez and Le Calvez (1958), Cimerman and Langer (1991), Levy *et al.* (1992), Sgarrella and Moncharmont-Zei (1993) and Hottinger *et al.* (1993).

Literature discussing modern and Quaternary foraminifers from North African coastal settings is limited in number. Studies include areas off Lebanon (Moncharmont Zei 1968) and Israel (Reinhardt *et al.* 1994; Yanko *et al.* 1994). Several investigations also focus on Nile delta sectors (El-Wakeel *et al.* 1970; Kulyk 1987; Abdou *et al.* 1991; Samir 2000; Samir and El Din 2001). For general taxonomy, Parker (1958) and Levy *et al.* (1992) are the most detailed references for the eastern Mediterranean. Interpretation of Aboukir foraminiferal associations examined here is established by comparison with ecological data from several Mediterranean areas. Particularly useful are those of the Gulf of Naples (Sgarrella and Moncharmont-Zei 1993), Venice Lagoon (Fiorini and Vaiani 2001), and the late Quaternary of the Po Plain near Ravenna (Albani and Serandrei Barbero 1990).

Table 4.2 Molluscan species identified in this study, with taxa listed in systematic order

BIVALVIA
Nucula nucleus Bronn
Nuculana fragilis (Chemnitz)
Arca noae L.
Acar scabra (Poli)
Anadara diluvii (Lamarck)
Striarca lactea (L.)
Glycymeris glycymeris (L.)
Mytilaster sp.
Musculus costulatus (Risso)
Modiolula phaseolina
Modiolus adriaticus (Lamarck)
Pinctada radiata (Leach)
Chlamys opercularis (L.)
Chlamys varia (L.)
Pododesmus squamulus (L.)
Lima lima (L.)
Ostrea edulis L.
Ostrea sp.
Ctena decussata (O.G. Costa)
Loripes lacteus (L.)
Chama gryphoides (L.)
Lepton nitidum Turton
Galeomma turtoni (Sowerby)
Glans aculeata (Poli)
Glans trapezia (L.)
Digitaria digitaria (L.)
Parvicardium exiguum (Gmelin in L.)
Plagiocardium sp.
Cerastoderma glaucum (Bruguière)
Mactra stultorum (L.)
Tellina pulchella Lamarck
Tellina sp.
Macoma cumana (O.G.Costa)
Macoma sp.
Donax semistriatus Poli
Donax trunculus L.
Donax sp.
Abra ovata (Philippi)
Pharus legumen (L.)
Corbicula fluminalis (Mueller)
Venus casina L.
Chamelea gallina (L.)
Gouldia minima (Montagu)
Dosinia lupinus (L.)
Venerupis aurea (Gmelin in L.)
Venerupis sp.
Petricola lithophaga (Retzius)
Corbula gibba (Olivi)
Hiatella arctica (L.)
Hiatella rugosa (Pennant)
Saxicavella plicata (Gmelin in Montagu)
Barnea candida (L.)
Thracia villosiuscula (McGilliwray)
GASTROPODA
Emarginula huzardi Payraudeau
Puncturella noachina (L.)
Acmaea virginea (Muller)
Cocculina sp.
Jujubinus sp.
Gibbula fanulum (Gmelin)
Gibbula sp.
Calliostoma zizyphinum (L.)
Calliostoma sp.
Tricolia speciosa (Von Muhlfeldt)
Smaragdia viridis (L.)
Cochliolepis costulatus (De Folis)
Teodoxus niloticus (Reeve)
Valvata nilotica
Hydrobia stagnalis (Baster)
Hydrobia sp.
Crisilla semistriata Montagu
Manzonia crassa (Kanmacher)
Setia sp.
Rissoa variabilis (Von Muhlfeldt)
Rissoa sp.
Turboella dolium (Nyst)
Turboella inconspicua (Alder)
Turboella lineolata (Michaud)
Alvania sp.
Vermetus triqueter (Bivona)
Parastrophia folini (L.)
Caecum trachaea (Montagu)
Melanoides tuberculata (Mueller)
Potamides conicus (Blainville)
Bittium reticulatum (Da Costa)
Cerithiopsis minima (Brusina)
Cerithiopsis tubercularis (Montagu)
Biforina perversa (Montagu)
Natica hebraea (Martyn)
Neverita josephinia Risso
Naticarius stercusmuscarum (Gmelin)
Muricopsis cristatus (Brocchi)
Columbella rustica (L.)
Cyclope neritea (L.)
Hinia pygmaea (Lamarck)
Hinia reticulata (L.)
Nassarius corniculus (Olivi)
Nassarius mutabilis (L.)
Conus sp.
Mangelia sp.
Ringicula conformis Monterosato
Chrysallida clathrata (Jeffreys)
Chrysallida fenestrata (Forbes in Jeffreys)
Chrysallida fischeri (Hornung & Mermod)
Chrysallida obtusa (Brown)
Chrysallida sp.
Chrysallida suturalis (Philippi)
Chrysallida terebellum (Philippi)
Chrysallida turbinelloides (Brusina)
Vitreolina curva (Monterosato)
Eulimella scillae (Scacchi)
Ebala nitidissima (Montagu)
Odostomia conoidea (Brocchi)
Turbonilla gradata B.D.D.
Turbonilla rufa (Philippi)
Turbonilla sp.
Biomphalaria alexandrina (Ehrenberg)
SCAPHOPODA
Dentalium dentalis L.

Table 4.3 Data matrix used for for the cluster analysis with molluscs.

Env. setting	*Species*	*ST 1-1*	*ST 1-2*	*ST 1-7*	*ST 1-8*	*ST2-1*	*ST2-2*	*ST2-3*	*ST2-4*	*ST2-5*	*ST3-1*	*ST3-2*	*ST3-3*	*ST3-4*
L (LEE excl)	*Abra ovata*	0	0	5	2	0	0	0	0	0	0	2	0	0
M	*Acar scabra*	0	0	0	0	0	0	0	0	1	1	0	1	0
M	*Acmaea virginea*	0	0	0	0	0	0	0	0	2	0	0	0	0
M	*Anadara diluvii*	0	0	0	0	0	0	0	0	1	0	0	0	0
M (AP pref)	*Arca noae*	0	0	0	0	0	0	0	0	0	0	0	0	0
M (SFBC)	*Barnea candida*	0	0	0	0	0	0	0	1	0	0	0	0	0
M (C)	*Biforina perversa*	0	0	0	0	1	0	0	0	0	0	0	0	0
FW	*Biomphalaria alexandrina*	0	0	0	0	0	0	0	0	0	0	0	0	0
M (AP-HP)	*Bittium reticulatum*	0	0	0	0	3	4	0	3	3	4	7	3	0
M	*Caecum trachaea*	0	0	0	0	0	1	0	0	0	0	0	0	0
M (DC)	*Calliostoma zizyphinum*	0	0	0	0	0	0	0	0	0	0	0	0	0
L (LEE excl)	*Cerastoderma glaucum*	0	0	7	6	0	0	0	0	0	0	0	0	0
M	*Cerithiopsis minima*	0	0	0	0	0	0	0	0	0	0	0	0	0
M (C)	*Cerithiopsis tubercularis*	0	0	0	0	0	0	1	0	0	0	2	0	0
M (AP-HP)	*Chama gryphoides*	0	0	0	0	0	0	0	0	0	0	0	0	0
M (SFBC pref)	*Chamelea gallina*	0	0	0	0	10	12	1	16	23	5	5	3	1
M	*Chlamys opercularis*	0	0	0	0	0	0	0	0	0	0	0	0	0
M	*Chlamys varia*	0	0	0	0	0	0	0	0	1	0	0	0	0
M	*Chrysallida clathrata*	0	0	0	0	0	0	0	0	0	0	0	0	0
M	*Chrysallida fenestrata*	0	0	0	0	0	0	0	0	0	0	0	0	0
M	*Chrysallida fischeri*	0	0	0	0	0	1	0	0	0	2	0	0	0
M	*Chrysallida obtusa*	0	0	0	0	0	0	0	0	0	0	0	0	0
M	*Chrysallida suturalis*	0	0	0	0	0	0	0	0	0	1	1	0	0
M	*Chrysallida terebellum*	0	0	0	0	0	0	0	0	0	0	0	0	0
M	*Chrysallida turbinelloides*	0	0	0	0	2	0	0	0	0	0	0	0	0
M	*Cochliolepis costulatus*	0	0	0	0	0	0	0	0	0	0	1	0	0
M (AP excl)	*Columbella rustica*	0	0	0	0	2	0	0	0	0	0	1	0	0
FW	*Corbicula fluminalis*	0	0	0	0	0	0	0	0	0	0	0	0	0
M (PE pref)	*Corbula gibba*	0	0	0	0	3	0	0	2	1	0	0	0	0
M	*Crisilla semistriata*	0	0	0	0	0	3	0	3	0	4	4	0	0
M (SVMC excl)	*Ctena decussata*	0	0	0	0	0	0	0	0	0	1	1	0	0
L (LEE excl)	*Cyclope neritea*	0	0	0	0	0	0	0	0	0	0	0	0	0
M	*Dentalium dentalis*	0	0	0	0	0	0	0	0	2	0	0	0	0
M	*Digitaria digitaria*	0	0	0	0	0	0	0	0	2	0	0	0	0
M (SFBC)	*Donax semistriatus*	0	0	0	0	5	4	2	1	17	2	6	0	1
M (SFHN excl)	*Donax trunculus*	0	0	0	0	0	0	0	0	2	0	0	0	0
M	*Dosinia lupinus*	0	0	0	0	0	0	0	0	0	0	7	0	0
M	*Ebala nitidissima*	0	0	0	0	0	0	0	0	0	0	0	1	0
M	*Emarginula huzardi*	0	0	0	0	0	0	0	0	1	0	0	0	0
M	*Eulimella scillae*	0	0	0	0	0	0	0	0	0	0	1	0	0
M	*Galeomma turtoni*	0	0	0	0	0	0	0	0	0	0	2	0	0
M (HP-DC)	*Gibbula fanulum*	0	0	0	0	0	0	0	1	2	0	0	0	0
M	*Glans aculeata*	0	0	0	0	0	0	0	0	0	1	0	0	0
M (HP excl)	*Glans trapezia*	0	0	0	0	0	0	0	0	0	0	0	0	0
M (SGCF excl)	*Glycymeris glycymeris*	0	0	0	0	58	44	1	34	14	11	26	4	0
M (DC)	*Gouldia minima*	0	0	0	0	0	0	0	0	0	0	0	0	0
M	*Hiatella arctica*	0	0	0	0	0	0	0	0	0	0	5	0	1
M	*Hiatella rugosa*	0	0	0	0	0	0	0	0	0	0	0	0	0
L (LEE excl)	*Hinia reticulata*	0	0	0	0	0	1	0	0	0	0	0	0	0
M (SFBC pref)	*Hinia pygmaea*	0	0	0	0	0	0	0	0	0	0	0	0	0
L (LEE excl)	*Hydrobia stagnalis*	0	0	1	1	0	0	0	0	0	0	3	0	0
M	*Lepton nitidum*	0	0	0	0	0	0	0	0	1	0	5	0	0
M	*Lima lima*	0	0	0	0	0	0	0	0	0	1	0	0	0
M (SVMC excl)	*Loripes lacteus*	1	0	0	0	0	0	0	0	7	1	1	0	0
M (SFBC)	*Macoma cumana*	0	0	0	0	0	0	0	0	4	1	5	2	0
M (SFBC)	*Mactra stultorum*	2	0	0	0	0	0	1	0	2	0	0	1	0
M	*Manzonia crassa*	0	0	0	0	0	0	0	3	0	0	0	0	0
FW	*Melanoides tuberculata*	0	0	0	0	0	0	0	0	0	0	0	0	0

ST7-1	*ST7-2*	*ST9-1*	*ST9-2*	*ST9-6*	*ST9-7*	*ST9-8*	*ST12-1*	*ST13-1*	*ST13-2*	*ST13-3*	*ST13-4*	*ST13-5*	*ST13-8*	*ST13-9*	*ST14-1*	*ST 17-1*	*ST 17-2*	*ST 17- 3*	*ST17-4*
0	0	1	0	3	4	79	0	1	0	0	0	3	25	185	4	11	75	64	106
0	0	0	0	0	0	0	0	0	0	0	0	0	0	0	0	0	0	0	0
0	0	0	0	0	0	0	0	0	0	0	0	0	0	0	0	0	0	0	0
0	0	0	0	0	0	0	0	0	0	0	0	0	0	0	0	1	0	0	0
0	0	0	0	0	0	0	0	0	0	1	0	0	0	0	0	0	0	0	0
0	0	0	0	0	0	1	0	0	0	0	0	0	0	1	0	0	0	0	0
0	0	1	0	0	0	0	0	0	0	0	0	0	0	0	0	0	0	0	0
0	0	0	1	0	0	0	0	0	0	0	0	0	0	0	0	0	0	0	0
1	1	11	12	0	1	7	2	8	4	2	0	0	4	19	6	15	0	7	2
0	0	0	0	0	0	0	0	0	0	0	0	0	0	0	1	2	0	0	0
0	0	0	1	0	0	0	0	0	1	0	0	0	0	0	0	0	0	0	0
0	0	0	1	2	7	81	0	1	0	0	0	1	34	105	11	0	31	58	186
0	0	0	2	0	0	0	0	2	0	0	0	0	0	0	0	0	0	0	0
0	0	0	3	0	0	0	0	0	0	0	0	0	0	0	0	0	0	0	0
0	0	1	0	0	0	0	0	0	0	0	0	0	0	0	0	0	0	0	0
3	2	2	0	0	0	1	1	2	0	0	1	0	0	0	4	3	0	2	1
0	0	0	0	0	0	0	0	0	1	0	0	0	0	0	0	0	0	0	0
0	0	0	0	0	0	0	0	0	0	0	0	0	0	0	0	0	0	0	0
0	0	0	0	0	0	0	0	0	0	0	0	0	0	2	0	0	0	0	0
0	0	0	0	0	0	0	0	0	0	0	0	0	0	1	0	0	0	0	0
0	0	0	0	0	0	0	0	0	0	0	0	0	0	0	0	0	0	0	0
0	0	1	0	0	0	2	0	0	0	0	0	0	0	0	0	0	0	4	64
0	0	0	0	0	0	3	0	2	1	0	0	0	0	0	0	0	0	0	0
0	0	0	0	0	0	0	0	0	0	0	0	0	0	4	0	0	0	0	0
0	0	0	0	0	0	0	0	0	0	0	0	0	0	0	0	0	0	0	0
0	0	0	1	0	0	0	0	0	0	0	0	0	0	0	0	0	0	0	0
0	0	0	0	0	0	0	0	0	0	0	0	0	0	0	0	0	0	0	0
0	0	0	0	0	2	6	0	0	0	0	0	0	0	9	0	0	0	0	0
0	0	0	0	0	0	1	0	1	0	0	0	0	0	0	0	0	0	1	0
0	3	1	1	0	0	0	1	1	0	0	0	0	0	0	1	0	0	0	0
0	0	0	0	0	0	3	0	1	0	0	0	0	0	0	0	0	0	0	0
0	0	0	0	0	0	0	0	0	0	0	0	0	0	0	1	0	1	9	1
0	0	0	0	0	0	0	0	0	0	0	0	0	0	0	0	0	0	0	0
0	0	0	0	0	0	0	0	0	0	0	0	0	0	0	0	3	0	0	0
1	0	1	0	0	1	1	0	0	0	0	0	0	0	0	1	0	0	0	0
0	0	0	0	0	0	0	0	0	0	0	0	0	0	0	1	0	0	0	0
0	0	0	0	0	0	0	0	0	0	1	0	0	0	0	0	0	0	0	0
0	0	0	0	0	0	0	0	0	0	0	0	0	0	0	0	0	0	0	0
0	0	0	0	0	0	0	0	0	0	0	0	0	0	0	0	0	0	0	0
0	0	0	0	0	0	0	0	0	0	0	0	0	0	0	0	0	0	0	0
0	0	0	0	0	0	0	0	0	0	0	0	0	0	0	0	0	0	0	0
0	0	0	0	0	0	0	0	1	0	0	0	0	0	0	0	0	0	0	0
0	0	0	0	0	0	0	0	2	2	0	0	0	0	0	0	0	0	0	0
0	0	0	4	0	0	0	0	0	0	0	0	0	0	0	0	0	0	0	0
6	2	6	0	0	0	2	2	4	3	0	0	0	1	1	7	11	0	0	0
0	0	0	0	0	0	0	0	1	0	0	0	0	0	0	0	0	0	0	0
0	0	1	0	0	0	0	0	0	0	0	0	0	0	0	0	0	0	0	0
0	0	0	0	0	0	0	0	1	0	0	0	0	0	0	0	0	0	0	0
0	0	2	1	0	0	2	0	0	0	0	0	0	0	0	0	0	0	0	0
0	0	0	0	0	0	0	0	1	3	0	0	0	0	0	0	0	0	0	0
0	0	1	0	0	11	148	0	0	0	0	0	0	185	816	30	22	384	447	410
0	0	2	2	0	0	0	0	0	2	0	0	0	0	0	3	0	0	0	1
0	0	1	0	0	0	0	0	0	0	0	0	0	0	0	0	0	0	0	0
0	0	0	0	0	0	10	0	0	0	0	0	0	4	23	3	6	0	16	6
1	0	0	0	0	0	0	1	0	0	0	0	0	0	2	4	0	0	0	0
0	0	0	0	0	0	0	0	0	0	0	0	0	0	3	0	0	0	1	0
0	0	1	3	0	0	0	0	2	0	0	0	0	0	0	0	2	0	0	0
0	0	0	0	0	0	2	0	0	0	0	0	0	3	10	0	0	0	0	0

Table 4.3 Data matrix used for for the cluster analysis with molluscs (continued).

Env. setting	*Species*	*ST 1-1*	*ST 1-2*	*ST 1-7*	*ST 1-8*	*ST2-1*	*ST2-2*	*ST2-3*	*ST2-4*	*ST2-5*	*ST3-1*	*ST3-2*	*ST3-3*	*ST3-4*
M (DC excl)	*Modiolula phaseolina*	0	0	0	0	0	0	0	0	0	0	0	0	0
M	*Modiolus adriaticus*	0	0	0	0	0	0	0	0	0	3	4	0	0
M (C pref)	*Muricopsis cristatus*	0	0	0	0	0	0	0	0	0	0	0	0	0
M	*Musculus costulatus*	1	1	3	3	3	3	0	6	28	12	36	12	6
M	*Nassarius corniculus*	0	0	0	0	0	0	0	0	1	0	0	0	0
M (SFBC pref)	*Nassarius mutabilis*	0	0	0	0	0	0	0	0	0	0	0	0	0
M	*Natica hebraea*	0	0	0	0	0	0	0	0	0	0	0	0	0
M	*Naticarius stercusmuscarum*	0	0	0	0	0	0	0	1	0	0	0	0	0
M (SFBC excl)	*Neverita josephinia*	0	0	0	0	1	0	0	0	0	1	0	0	0
M	*Nucula nucleus*	0	0	0	0	1	0	0	0	0	0	1	0	0
M	*Nuculana fragilis*	0	0	0	0	0	0	0	0	0	0	0	0	0
M	*Odostomia conoidea*	0	0	0	0	1	1	0	0	0	0	0	0	0
M	*Ostrea edulis*	0	0	0	0	1	0	0	0	2	0	0	0	0
M	*Parastrophia folini*	0	0	0	0	0	0	0	0	1	0	0	0	0
M	*Parvicardium exiguum*	0	0	0	0	0	0	0	0	0	0	0	0	0
M	*Petricola lithophaga*	0	0	0	0	0	0	0	0	0	0	0	0	0
M	*Pharus legumen*	0	0	0	0	0	0	0	0	1	0	0	0	0
M	*Pinctada radiata*	0	0	0	0	0	0	0	0	6	2	3	0	0
M	*Pododesmus squamulus*	0	0	0	0	0	0	0	0	6	0	1	0	0
L (LEE)	*Potamides conicus*	0	1	0	0	0	3	0	0	0	0	0	0	0
M	*Puncturella noachina*	0	0	0	0	0	0	0	0	1	0	0	0	0
M	*Ringicula conformis*	0	0	0	0	2	3	0	0	0	1	1	0	0
M	*Rissoa variabilis*	0	0	0	0	0	0	0	0	0	0	1	0	0
M	*Saxicavella plicata*	0	0	0	0	0	0	0	0	6	0	0	0	0
M (AP-HP)	*Smaragdia viridis*	0	0	0	0	1	0	0	0	0	0	0	0	0
M	*Striarca lactea*	0	0	0	0	0	0	0	0	1	0	0	0	0
M (SFBC excl)	*Tellina pulchella*	0	0	0	0	0	0	0	0	0	1	4	0	0
FW	*Teodoxus niloticus*	0	0	0	0	0	0	0	0	0	0	0	0	0
M	*Thracia villosiuscula*	0	0	0	0	0	0	0	0	0	0	0	0	0
M	*Tricolia speciosa*	0	0	0	0	6	0	1	1	1	5	2	0	1
M (HP)	*Turboella dolium*	0	0	0	0	0	0	0	0	0	0	0	0	0
M	*Turboella inconspicua*	0	0	0	0	0	0	0	0	0	0	0	0	0
M	*Turboella lineolata*	0	0	0	0	0	0	0	0	0	0	0	0	0
M	*Turbonilla gradata*	0	0	0	0	0	0	0	0	0	0	0	0	0
M (PE pref)	*Turbonilla rufa*	0	0	0	0	0	0	0	0	0	0	0	0	0
FW	*Valvata nilotica*	0	0	0	0	0	0	0	0	0	0	0	0	0
M (SVMC excl)	*Venerupis aurea*	0	0	0	0	0	0	0	0	0	0	0	0	0
M (SGCF excl)	*Venus casina*	0	0	0	0	0	0	0	0	0	0	0	0	0
M	*Vermetus triqueter*	0	0	0	0	0	0	0	0	6	0	2	2	0
M	*Vitreolina curva*	0	0	0	0	0	0	0	0	0	0	2	0	0

The foraminifers found in the 78 core samples account for 125 species, and are listed in alphabetical order in Table 4.4. Most are benthic and taxa are well known in the modern Mediterranean. Some species, such as *Cibicides tabaensis, Pseudotriloculina philippiensis, Spiroloculina antillarum* and *W. auriculata,* are described in the Gulf of Aqaba, Red Sea (Hottinger *et al.* 1993). Sixteen (20%) of the 78 Aboukir Bay core samples did not contain any foraminifers (denoted by the letter n; Figures 4.2 and 4.3).

The database for cluster analysis was obtained retaining samples with at least two specimens; in this matrix, the abundance value for each taxon in each sample is reported (Table 4.5). Five groups are recognised: **brackish water, brackish water–infralittoral, infralittoral** and **infralittoral– circalittoral**. Infralittoral and circalittoral environments correspond to marine environment settings. Taxa in open nomenclature (sp., cf., aff.) show morphological features similar to modern species, and thus allow these latter to be included in one of the groups. It is noted that some samples contain iron-stained tests (Figures 4.2, 4.3; see Discussion section). When they occur in a sample together with well-preserved taxa, the oxidised specimens are considered

ST7-1	*ST7-2*	*ST9-1*	*ST9-2*	*ST9-6*	*ST9-7*	*ST9-8*	*ST12-1*	*ST13-1*	*ST13-2*	*ST13-3*	*ST13-4*	*ST13-5*	*ST13-8*	*ST13-9*	*ST14-1*	*ST 17-1*	*ST 17-2*	*ST 17- 3*	*ST17-4*
0	0	0	0	0	0	0	0	0	1	0	0	0	0	0	0	5	0	1	0
3	0	13	23	0	0	1	0	6	1	2	0	0	0	4	0	0	0	0	2
0	0	0	1	0	0	0	0	0	0	0	0	0	0	0	0	0	0	0	0
0	0	11	25	0	0	1	0	7	1	2	0	0	0	0	16	12	0	0	2
0	0	1	0	0	0	0	0	0	0	0	0	0	0	0	0	0	0	0	0
0	0	0	0	0	0	0	0	0	0	0	0	0	0	0	0	0	0	1	0
0	0	1	1	0	0	0	0	0	0	0	0	0	0	0	0	0	0	0	0
0	0	0	0	0	0	0	0	0	0	0	0	0	0	0	0	0	0	0	0
0	0	0	0	0	0	0	0	0	0	0	0	0	0	0	0	0	0	4	0
0	0	0	0	0	0	0	0	0	0	1	0	0	0	0	0	0	0	0	0
0	0	0	0	0	0	1	0	0	0	0	0	0	0	0	0	0	0	0	0
0	0	3	1	0	0	0	0	2	1	0	0	0	0	1	0	0	0	1	0
0	0	1	0	0	0	0	0	0	0	0	0	0	0	0	6	2	0	1	0
0	0	0	0	0	0	0	0	0	0	0	0	0	0	0	0	0	0	0	0
0	0	0	1	0	0	0	0	0	0	0	0	0	0	0	0	0	0	0	0
0	0	2	0	0	0	0	0	0	0	0	0	0	0	0	0	0	0	0	0
0	0	0	0	0	0	0	0	0	1	0	0	0	0	0	0	0	0	0	0
0	0	1	3	0	0	0	0	0	0	0	0	0	0	0	1	1	0	0	0
0	0	0	3	0	0	0	0	0	0	0	0	0	0	0	0	0	0	0	0
0	0	0	0	0	0	44	0	0	1	0	0	0	0	0	2	0	10	50	19
1	0	0	0	0	0	0	0	0	0	0	0	0	0	0	0	0	0	0	0
0	0	1	1	0	0	0	0	2	0	0	0	0	0	0	0	0	0	0	0
0	0	0	3	0	0	3	0	0	0	0	0	0	0	0	0	0	0	0	0
0	0	0	0	0	0	0	0	0	0	0	0	0	0	0	0	0	0	0	0
0	0	0	0	0	0	1	0	0	0	0	0	0	0	0	0	0	0	0	0
0	0	0	0	0	0	0	0	0	0	0	0	0	0	0	0	0	0	0	0
0	0	0	0	0	0	0	0	0	0	0	0	0	0	0	0	0	0	0	0
0	0	0	0	0	0	0	0	0	0	0	0	0	0	2	0	0	0	0	0
0	0	0	0	0	0	0	0	1	1	0	0	0	0	0	0	1	0	0	0
0	0	5	15	0	0	0	0	1	1	1	2	0	0	2	3	3	0	0	0
0	0	0	0	0	0	0	0	0	0	0	0	0	0	0	0	0	3	9	0
0	0	0	0	0	0	0	0	1	0	0	0	0	0	0	0	0	0	0	0
0	0	0	0	0	0	0	0	0	0	0	0	0	0	15	0	0	0	0	0
0	0	0	1	0	0	0	0	0	0	0	0	0	0	0	0	0	0	0	0
0	1	0	0	0	0	0	0	0	0	0	0	0	0	0	0	0	0	0	0
0	0	0	0	0	0	0	0	0	0	0	0	0	0	2	0	0	0	0	0
0	0	5	3	0	0	0	0	0	0	0	0	0	0	5	0	0	0	7	0
0	0	2	1	0	0	0	0	0	0	0	0	0	0	0	0	1	0	0	0
0	0	0	0	0	0	0	0	0	0	0	0	0	0	0	0	0	0	0	0
0	0	0	0	0	0	0	0	0	0	0	0	0	0	0	0	0	0	0	0

allochthonous (i.e. having been reworked and displaced). In contrast, when oxidised tests occur in samples where all the specimens are stained, they are considered autochthonous (i.e. having lived in place).

2.4 Ostracods

Ostracods, unlike foraminifers, were picked directly from bulk samples and then identified. The number of specimens was reported for each sample, and the presence of juveniles recorded. Selected specimens were examined and photographed by a Leica Scanning Electron Microscope at the University of Trieste. There is an abundant literature pertaining to modern and Quaternary ostracods on the northern African margin. Modern marine coastal ostracods in the near-shore of Tripoli (Libya), Bay of Bou-Ismail (Algeria) and Tunisian shelf have been studied (respectively, Bonaduce and Pugliese 1975; Yassini 1979; Bonaduce *et al.* 1988). Moreover, brackish water ostracods of Ghar el Melah lagoon (Tunisia), microfaunae of Tunis Lake, and ostracods of Aboukir Bay and Edkou and Manzala lagoons have also been investigated (among other authors, Mansouri and Carbonel

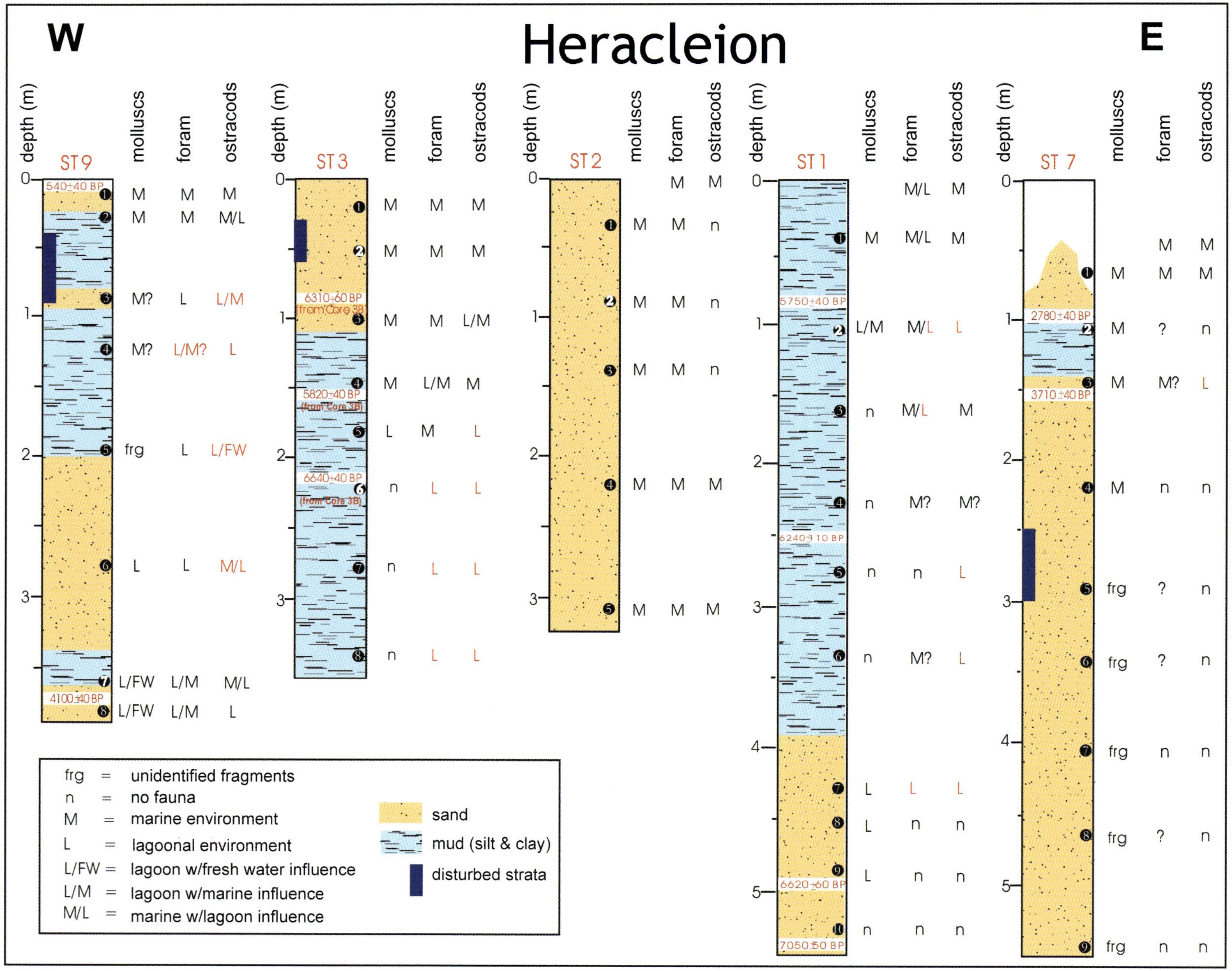

Figure 4.2 Simplified lithological logs recovered at Heracleion showing samples examined in this study (numbers in black circle) for fauna and grain size. Letters represent non-oxidised (in black) and oxidised (in red) biofacies. Radiocarbon dates are not calibrated. The environmental interpretations derived from the fauna are shown. Those denoted with (?) record poor fauna.

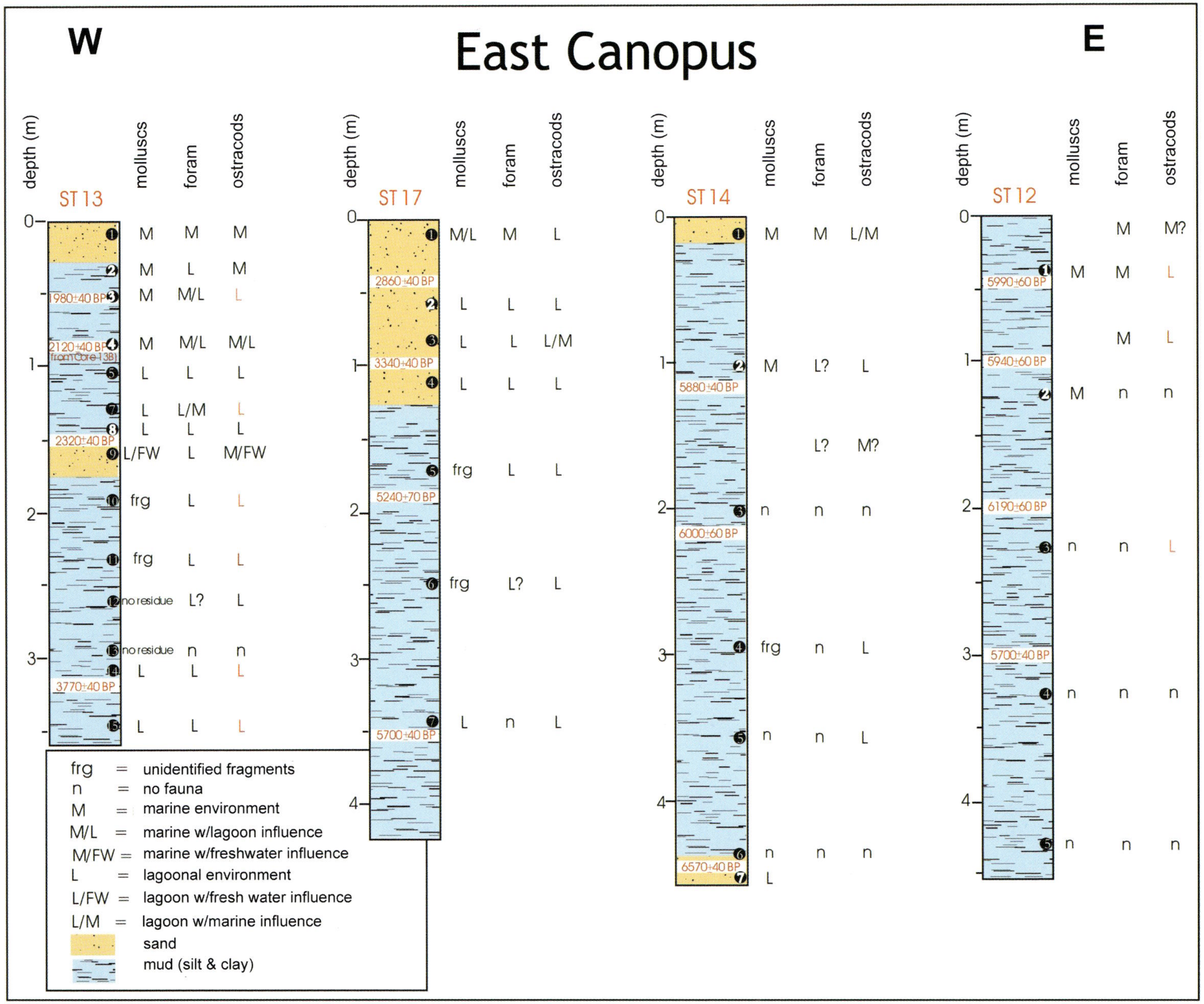

Figure 4.3 Simplified lithological logs recovered at East Canopus showing samples examined in this study (numbers in black circle) for fauna and grain size. Coloured letter symbols represent non-oxidised (in black) and oxidised (in red) biofacies. Radiocarbon dates are not calibrated. The environmental interpretations derived from the fauna are indicated. Those denoted with (?) record poor fauna.

Table 4.4 Foraminifer species identified in this study, with taxa listed in alphabetical order

Adelosina carinata-striata Wiesner, 1923
Adelosina cliarensis (Heron-Allen & Earland, 1930)
Adelosina dubia (d'Orbigny, 1826)
Adelosina longirostra (d'Orbigny, 1826)
Adelosina mediterranensis (Le Calvez J. & Y., 1958)
Adelosina pulchella d'Orbigny, 1846
Affinetrina bermudezi (Acosta, 1940)
Affinetrina eburnea (d'Orbigny, 1839)
Affinetrina ucrainica (Serova, 1952)
Ammonia beccarii (Linneo, 1758)
Ammonia papillosa (d'Orbigny, 1850)
Ammonia parkinsoniana (d'Orbigny, 1839)
Ammonia tepida (Cushman, 1926)
Amphistegina madagascariensis d'Orbigny, 1826
Articulina mucronata (d'Orbigny, 1839)
Asterigerinata mamilla (Williamson, 1858)
Astrononion stelligerum (d'Orbigny, 1839)
Biloculinella labiata (Schlumberger, 1891)
Bolivina spp.
Buccella granulata (Di Napoli Alliata, 1952)
Bulimina elongata d'Orbigny, 1846
Cibicidella variabilis (d'Orbigny, 1826)
Cibicides lobatulus (Walker & Jacob, 1798)
Cibicides refulgens Monfort, 1808
Cibicides sp.1
Cibicides tabaensis Perelis & Reiss, 1975
Coscinospira hemprichii Ehremberg, 1839
Cycloforina aff. *juleana* (d'Orbigny, 1846)
Cycloforina costata (d'Orbigny, 1826)
Cycloforina juleana (d'Orbigny, 1846)
Cycloforina schlumbergeri (Wiesner, 1923)
Cycloforina sp.1
Cycloforina tenuicollis (Wiesner, 1923)
Cymbaloporetta bradi (Cushman, 1928)
Discorbinella bertheloti (d'Orbigny, 1839)
Elphidium aculeatum (d'Orbigny, 1846)
Elphidium crispum (Linneo, 1758)
Elphidium excavatum (Terquem, 1876)
Elphidium gerthi van Voorthuysen, 1957
Elphidium granosum (d'Orbigny, 1846)
Elphidium jenseni (Cushman, 1924)
Elphidium macellum (Fichtel & Moll, 1798)
Elphidium punctatum (Terquem, 1878)
Eponides pusillus Parr, 1950
Eponides repandus (Fichtel & Moll, 1798)
Fischerina compressa (Wiesner, 1923)
Genus A
Glabratella erecta (Sidebottom, 1908)
Glabratella exacamerata Sigle & Bermudez, 1965
Glabratella torrei (Bermudez, 1935)
Heterostegina antillarum d'Orbigny, 1826
Lachlanella reticolata (d'Orbigny, 1826)
Lamarkina scabra (Brady, 1884)
Lenticulina sp.
Massilina gualtieriana (d'Orbigny, 1839)
Massilina secans (d'Orbigny, 1826)
Miliolinella dilatata (d'Orbigny, 1839)
Miliolinella elongata Kruit, 1955
Miliolinella labiosa (d'Orbigny, 1839)
Miliolinella semicostata (Wiesner, 1923)
Miliolinella sp.1
Miliolinella webbiana (d'Orbigny, 1839)
Parrina bradyi (Millett, 1898)
Peneroplis pertusus (Forskål, 1775)
Peneroplis planatus (Fichtel & Moll, 1798)
Planoglabratella opercularis (d'Orbigny, 1839)
Planorbulina mediterranensis d'Orbigny, 1826
Polymorphyna spp.
Pseudoeponides sp.
Pseudotriloculina laevigata (d'Orbigny, 1826)
Pseudotriloculina philippiensis (Cushman, 1921)
Pyrgo oblunga (d'Orbigny, 1839)
Quinqueloculina aff. *stalkeri* Loebich & Tappan, 1953
Quinqueloculina berthelotiana d'Orbigny, 1839
Quinqueloculina bosciana d'Orbigny, 1839
Quinqueloculina cf. subpolygona Parr, 1945
Quinqueloculina lata Terquem, 1876
Quinqueloculina milletti (Wiesner, 1812)
Quinqueloculina nodulosa Wiesner, 1923
Quinqueloculina parvula Schlumberger 1894
Quinqueloculina peseudobuchiana Luczkowska, 1974
Quinqueloculina seminulum (Linneo, 1758)
Quinqueloculina sp.1
Quinqueloculina sp.2
Quinqueloculina sp.3
Quinqueloculina trigonula Terquem, 1876
Quinqueloculina ungeriana Terquem, 1876
Quinqueloculina vulgaris d'Orbigny, 1826
Rectoglandulina sp.
Reussella spinulosa (Reuss, 1850)
Rosalina bradyi (Cushman, 1915)
Rosalina floridana (Cushman, 1922)
Rosalina globularis d'Orbigny, 1826
Schackoinella imperatoria (d'Orbigny, 1846)
Sigmoilina costata Schlumberger, 1893
Sinuloculina consobrina (d'Orbigny, 1846)
Sinuloculina ciclostoma (Reuss, 1850)
Sinuloculina inflata (d'Orbigny, 1846)
Sinuloculina rotonda (d'Orbigny, 1893)
Siphonaperta agglutinans
Siphonaperta annectens (Schlumberger, 1893)
Siphonaperta aspera (d'Orbigny, 1826)
Siphonaperta irregularis (d'Orbigny, 1826)
Siphonaperta torrei (Acosta, 1939)
Siphonina reticolata (Czjzek, 1848)
Sorites orbiculus Ehrenberg, 1839
Spiroloculina affixa Terquem, 1878
Spiroloculina antillarum d'Orbigny, 1839
Spiroloculina canaliculata d'Orbigny, 1846
Spiroloculina cf. *dilatata* d'Orbigny, 1846
Spiroloculina excavata d'Orbigny, 1846
Spiroloculina ornata tricarinata Le Calvez J. & Y., 1858
Spirophthalmidium sp.
Textularia bocki Höglund, 1947
Textularia calva Lalicker, 1935
Textularia conica d'Orbigny, 1839
Textularia sp.1
Triloculina marioni Schlumberger, 1893
Triloculina placata Terquem, 1878
Triloculina sp.1
Triloculina tricarinata d'Orbigny, 1826
Triloculina trigonula (Lamarck, 1804)
Uvigerina sp.
Vertebralina striata d'Orbigny, 1826
Wiesnerella auriculata (Egger, 1895)

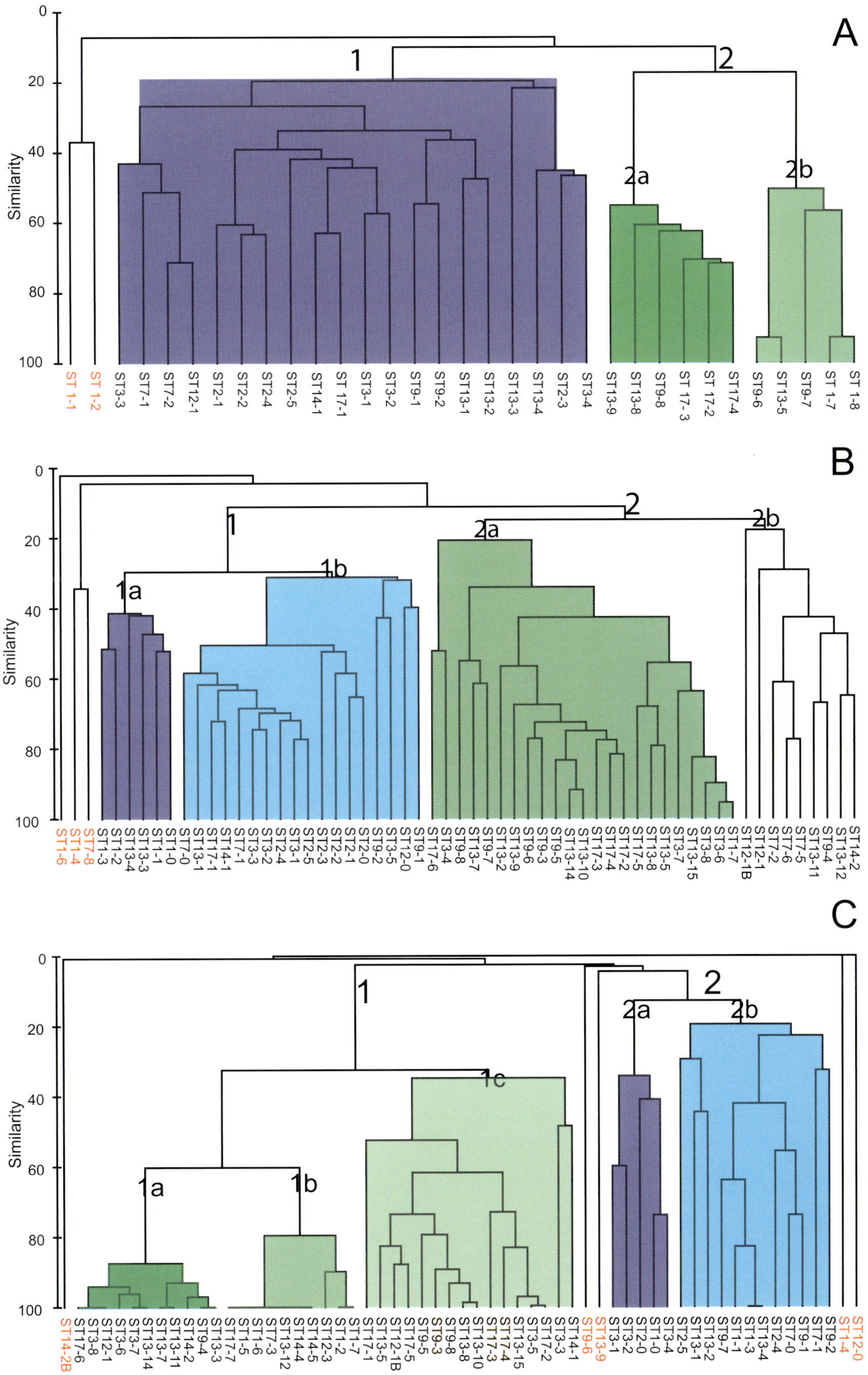

Figure 4.4 Cluster analysis plots for (A) molluscs, (B) foraminifera and (C) ostracods. Clusters in various blue colours group samples denoting marine environments; those in various green shades group samples of several lagoonal sub-environments. Samples numbered in red are unclustered.

Table 4.5 Data matrix used for the cluster analysis with foraminifers.

Env. setting	*Species*	*ST1-0*	*ST1-1*	*ST1-2*	*ST1-3*	*ST1-4*	*ST1-6*	*ST1-7*	*ST2-0*	*ST2-1*	*ST2-2*	*ST2-3*	*ST2-4*	*ST2-5*	*ST3-1*	*ST3-2*	*ST3-3*	*ST3-4*	*ST3-5*	*ST3-6*	*ST3-7*	*ST3-8*	*ST7-0*	*ST7-1*	*ST7-2*	*ST7-5*
M	*Adelosina cliarensis*	0	0	0	0	0	0	0	1	1	1	2	2	3	2	1	0	0	0	0	0	0	0	2	0	0
M	*Adelosina dubia*	0	0	1	3	0	0	0	3	2	3	3	10	7	8	4	5	0	0	0	0	0	5	5	0	0
M	*Adelosina longirostra*	0	0	0	0	0	0	0	1	1	1	1	3	0	0	1	4	0	0	0	0	0	6	2	0	0
M	*Adelosina mediterranensis*	0	0	0	0	0	0	0	0	0	0	0	0	0	0	0	0	0	0	0	0	0	0	0	0	0
M	*Affinetrina bermudezi*	0	1	1	0	0	0	0	0	2	1	0	0	0	1	4	1	0	0	0	0	0	1	3	0	0
M	*Affinetrina ucrainica*	6	0	1	0	0	1	0	1	2	0	1	6	4	7	3	13	0	1	0	0	0	6	2	0	0
M	*Ammomia beccarii*	1	0	0	0	0	0	0	0	0	0	0	0	0	0	0	0	0	0	0	0	0	0	0	0	0
M/L	*Ammonia papillosa*	0	0	0	0	0	0	0	0	0	1	0	0	0	0	0	0	0	0	0	0	0	0	0	0	0
L/M	*Ammonia parkinsoniana*	5	3	1	0	0	0	0	0	0	2	0	1	1	0	1	4	0	0	0	0	0	3	3	2	2
L	*Ammonia tepida*	7	2	7	9	0	0	49	0	0	0	0	5	4	13	17	16	4	13	41	10	30	3	2	0	0
M	*Amphistegina lobifera*	0	0	0	0	0	0	0	0	0	0	0	0	0	0	0	0	0	0	0	0	0	0	0	0	0
M	*Articulina mucronata*	0	0	0	0	0	0	0	0	0	0	0	0	2	1	0	0	0	0	0	0	0	0	0	0	0
M	*Asterigerinata mamilla*	0	0	1	1	0	0	0	0	0	0	2	2	3	3	2	1	0	2	0	0	0	9	4	0	0
M	*Biloculinella labiata*	0	0	0	0	0	1	0	0	0	0	0	0	0	0	0	0	0	0	0	0	0	0	0	0	0
M	*Bolivina* spp.	0	0	0	0	0	0	0	0	0	1	2	11	3	8	4	0	0	0	0	0	0	2	7	0	0
M/L	*Buccella granulata*	0	0	0	0	0	0	0	1	0	0	0	2	0	0	0	1	0	1	0	0	0	1	2	0	0
M	*Cibicidella variabilis*	0	0	0	0	0	0	0	0	0	0	2	3	0	2	1	4	0	0	0	0	0	2	0	0	0
M	*Cibicides lobatulus*	1	2	0	0	0	0	0	1	1	2	2	4	6	6	4	7	0	1	0	0	0	6	9	2	0
M	*Cibicides refulgens*	0	1	0	0	0	0	0	0	0	2	0	0	0	0	1	0	0	0	0	0	0	0	1	0	0
M	*Cibicides* sp. 1	0	0	0	0	0	0	0	4	3	0	2	9	0	1	2	2	0	1	0	0	0	2	6	0	0
M	*Cycloforina* aff. *juleana*	1	0	0	0	0	0	0	0	0	0	1	8	3	2	3	3	0	0	0	0	0	0	2	0	0
M/L	*Cycloforina costata*	2	1	2	0	0	0	0	1	4	2	1	4	5	10	9	4	0	0	0	0	0	1	5	0	0
M	*Cycloforina juleana*	0	1	1	0	0	0	0	0	0	0	0	5	3	0	0	0	0	0	0	0	0	0	0	0	0
M	*Cycloforina rugosa*	0	0	0	0	0	0	0	0	0	1	0	0	0	1	0	0	0	1	0	0	0	0	0	0	0
M	*Cycloforina schlumbergeri*	0	0	3	2	0	0	0	2	2	1	7	15	10	14	12	13	0	0	0	0	0	7	13	0	0
M	*Cycloforina tenuicollis*	1	1	0	2	0	0	0	0	0	0	0	1	1	5	7	2	0	0	0	0	0	1	3	0	0
M	*Cymbaloporetta bradyi*	0	0	0	0	0	0	0	1	0	1	0	0	1	0	0	0	0	0	0	0	0	1	0	0	0
M	*Discorbis* sp.	0	0	0	0	0	0	0	0	0	0	0	1	0	0	2	4	0	0	0	0	0	0	3	0	0
M	*Elphidium crispum*	0	3	0	0	0	0	0	0	0	0	0	0	0	0	0	0	0	0	0	0	0	0	0	0	0
M/L	*Elphidium excavatum*	0	0	0	0	0	0	0	0	0	1	0	0	0	0	0	1	0	0	0	0	0	0	0	0	0
M	*Elphidium gerthi*	1	0	0	0	0	0	0	1	1	0	0	3	5	8	6	7	0	0	0	0	0	7	9	0	0
M/L	*Elphidium granosum*	0	0	0	0	0	0	0	0	0	0	0	0	0	0	1	0	0	0	0	0	0	0	2	0	0
M	*Elphidium jenseni*	1	1	1	0	0	0	0	3	1	2	0	5	5	11	6	7	0	0	0	0	0	1	6	0	0
M	*Elphidium macellum*	0	2	0	0	1	0	0	1	1	2	0	1	3	0	6	4	0	0	0	0	0	0	2	0	0
M	*Elphidium punctatum*	0	2	1	0	0	0	0	0	0	2	0	1	0	0	4	2	0	1	0	0	0	0	0	0	0
M	*Eponides repandus*	0	1	0	0	0	0	0	2	1	1	0	0	1	1	3	0	0	0	0	0	0	1	0	0	0
M	*Glabratella* spp.	0	0	0	0	0	0	0	0	0	1	3	4	2	4	4	0	0	0	0	0	0	0	5	0	0
M	*Lachlanella reticulata*	0	0	0	0	0	0	0	1	0	0	0	1	0	0	0	0	0	0	0	0	0	0	1	0	0
M	*Lamarkina scabra*	0	0	0	0	0	0	0	0	0	0	0	0	0	0	1	1	0	0	0	0	0	0	0	0	0
M	*Lenticulina* sp.	0	0	0	0	0	0	0	0	0	0	0	0	0	2	0	1	0	0	0	0	0	0	4	0	0
M	*Massilina gualtieriana*	2	5	1	4	0	0	0	18	10	12	8	14	14	25	15	19	0	1	0	0	0	10	7	0	0
M	*Massilina secans*	0	0	0	0	0	0	0	3	2	4	0	7	5	3	3	4	0	0	0	0	0	0	0	0	0
M	*Miliolinella dilatata*	0	0	0	0	0	0	0	1	0	0	0	2	7	7	12	8	0	2	0	0	0	1	0	0	0
M	*Miliolinella elongata*	0	0	0	0	0	0	0	2	0	0	0	1	0	2	6	2	0	0	0	0	0	1	2	0	0
M	*Miliolinella labiosa*	0	0	0	0	0	0	0	0	0	1	0	13	5	3	3	3	0	0	0	0	0	0	6	0	0
M	*Miliolinella* sp.1	0	0	0	0	0	0	0	0	0	0	0	0	0	0	6	2	0	0	0	0	0	3	1	0	0
M	*Miliolinella webbiana*	0	0	1	0	0	0	0	0	0	0	0	0	1	0	0	2	0	0	0	0	0	0	0	0	0
M	*Nonion stelligerum*	0	0	0	0	0	0	0	0	0	3	2	2	12	8	13	8	0	0	0	0	0	4	7	0	0
M	*Parrina bradyi*	0	0	0	0	0	0	0	0	0	0	0	0	1	0	1	1	0	0	0	0	0	0	0	0	0
M	*Peneroplis pertusus*	7	7	6	3	0	0	0	7	1	5	2	8	5	7	8	2	0	0	0	0	0	11	24	0	0
M	*Peneroplis planatus*	5	0	1	0	0	0	0	4	2	6	4	6	7	5	4	1	0	3	0	0	0	0	1	0	0
M	*Planoglabratella opercularis*	0	0	0	0	0	0	0	1	0	1	0	0	0	0	1	0	0	0	0	0	0	0	0	0	0
M	*Planorbulina mediterranensis*	0	1	0	0	0	0	0	4	2	2	0	10	6	8	9	4	0	0	0	0	0	2	11	0	0
M	*Polymorphyna* spp.	0	0	0	0	0	0	0	4	3	7	4	5	1	7	4	4	0	1	0	0	0	0	0	0	0
M	*Pseudotriloculina laevigata*	0	0	0	0	0	0	0	0	0	0	0	1	3	0	0	2	0	0	0	0	0	0	0	0	0
M	*Quinqueloculina parvula*	0	0	0	0	0	0	0	0	0	0	0	5	2	0	0	0	0	0	0	0	0	0	0	0	0

ST7-6	*ST7-8*	*ST9-1*	*ST9-2*	*ST9-3*	*ST9-4*	*ST9-5*	*ST9-6*	*ST9-7*	*ST9-8*	*ST12-0*	*ST12-1*	*ST12-1B*	*ST13-1*	*ST13-2*	*ST13-3*	*ST13-4*	*ST13-5*	*ST13-7*	*ST13-8*	*ST13-9*	*ST13-10*	*ST13-11*	*ST13-12*	*ST13-14*	*ST13-15*	*ST14-1*	*ST14-2*	*ST17-1*	*ST17-2*	*ST17-3*	*ST17-4*	*ST17-5*	*ST17-6*
0	0	2	0	1	0	0	0	0	0	0	2	0	2	0	0	0	0	0	0	0	0	0	0	0	0	6	0	1	0	0	0	0	0
0	0	0	0	0	0	0	0	0	0	0	0	0	9	1	0	0	0	0	0	2	0	0	0	0	0	6	0	3	0	0	0	0	0
0	0	0	1	0	0	0	0	0	0	1	0	0	0	0	0	0	0	0	0	0	0	0	0	0	0	2	0	2	0	0	0	0	0
0	0	0	0	0	0	0	0	0	0	0	0	1	1	0	0	0	0	0	0	0	0	0	0	0	0	2	0	2	0	0	0	0	0
0	0	0	1	0	0	0	0	0	0	0	0	0	1	0	0	0	0	0	0	0	0	0	0	0	0	2	0	2	0	0	0	0	0
0	0	0	0	0	0	0	0	0	0	0	0	0	3	1	0	2	0	0	0	0	0	0	0	0	0	3	0	2	0	0	0	0	0
0	0	0	0	0	0	0	0	0	1	0	0	0	0	0	0	0	0	0	0	0	0	0	0	0	0	0	0	0	0	0	0	0	0
0	0	0	2	0	0	0	0	0	0	0	0	1	0	0	0	0	0	0	0	0	0	0	0	0	0	1	0	0	0	0	0	0	0
5	0	6	0	0	3	0	2	22	23	0	5	1	0	2	3	7	11	15	39	7	0	6	1	0	0	6	2	5	0	1	0	1	0
0	0	1	8	345	6	204	574	16	31	0	0	0	20	213	4	38	149	25	78	452	294	8	1	290	43	25	1	82	451	281	396	70	2
0	0	16	15	0	0	0	0	0	0	0	0	0	0	0	0	3	0	0	0	0	0	0	0	0	0	0	0	0	0	0	0	0	0
0	0	0	0	0	0	0	0	0	0	0	0	0	0	0	0	0	0	0	0	0	0	0	0	0	0	0	0	0	0	0	0	0	0
0	0	4	0	0	0	0	0	0	0	1	0	0	4	0	0	1	0	0	0	0	0	0	0	0	0	2	0	1	0	0	0	0	0
0	0	0	0	0	0	0	0	0	0	0	0	2	1	0	0	0	0	0	0	0	0	0	0	0	0	0	0	0	0	0	0	0	0
0	0	0	0	0	0	0	0	0	0	0	0	0	3	0	0	0	0	0	0	0	0	0	0	0	0	3	0	3	0	0	0	0	0
0	0	0	3	0	0	0	0	0	2	1	0	0	0	0	1	1	0	0	0	0	0	0	0	0	0	1	0	0	0	0	0	0	0
0	0	0	0	0	0	0	0	0	0	0	0	0	2	0	0	0	0	0	0	0	0	0	0	0	0	0	0	0	0	0	0	0	0
0	2	0	0	1	0	0	0	0	0	1	1	0	3	1	1	0	0	0	0	0	0	0	0	0	0	5	0	2	0	0	0	0	0
0	0	0	0	0	0	0	0	0	0	0	0	0	0	0	0	0	0	0	0	0	0	0	0	0	0	0	0	0	0	0	0	0	0
0	0	0	2	0	0	0	0	0	3	0	0	0	0	0	0	0	0	0	0	0	0	0	0	0	0	0	0	0	0	0	0	0	0
0	0	0	0	0	0	0	0	0	0	0	0	0	1	0	0	0	0	0	0	0	0	0	0	0	0	1	0	5	0	0	0	0	0
0	0	0	1	0	0	0	0	0	0	1	0	0	2	0	0	1	0	0	0	1	0	0	0	0	0	2	0	2	1	0	0	0	0
0	0	0	1	0	0	0	0	0	0	0	0	0	1	0	0	0	0	0	0	0	0	0	0	0	0	0	0	0	0	0	0	0	0
0	0	0	1	0	0	0	0	0	0	0	1	0	0	0	0	0	0	0	0	0	0	0	0	0	0	0	0	0	0	0	0	0	0
0	0	0	0	1	0	0	0	0	0	0	0	0	4	0	0	1	0	0	0	0	0	0	0	0	0	5	0	7	0	0	0	0	0
0	0	0	0	0	0	0	0	0	0	1	0	0	1	0	0	0	0	0	0	0	0	0	0	0	0	1	0	3	0	0	0	0	0
0	0	0	0	0	0	0	0	0	0	0	0	0	0	0	0	0	0	0	0	0	0	0	0	0	0	3	0	3	0	0	0	0	0
0	0	0	0	0	0	0	0	0	0	0	0	0	4	0	0	1	0	0	0	0	0	0	0	0	0	0	0	0	0	0	0	0	0
0	0	0	1	0	0	0	0	2	0	0	0	0	2	0	0	0	0	1	0	0	0	0	0	0	0	1	0	0	5	4	0	8	0
0	0	0	0	16	0	2	8	2	0	0	0	0	1	4	0	0	0	0	0	1	1	0	0	7	1	1	0	0	0	0	0	0	0
0	0	0	0	0	0	0	0	0	0	1	0	0	5	0	0	0	0	0	0	0	0	0	0	0	0	7	0	7	0	0	0	0	0
0	0	0	0	5	1	1	5	0	0	0	0	0	0	0	0	1	0	0	0	2	2	0	0	0	0	0	0	0	0	0	4	0	0
0	0	0	0	0	0	0	0	2	10	0	0	0	0	1	0	0	0	1	0	1	0	0	0	0	0	12	0	3	0	0	0	0	0
0	2	0	0	0	0	0	0	3	15	0	0	0	1	1	2	0	0	0	0	3	0	0	0	0	0	0	0	0	0	0	0	0	0
0	0	0	2	0	0	0	0	1	7	0	0	0	0	0	1	1	0	0	0	1	0	0	0	0	0	2	0	0	1	0	0	0	0
0	0	1	0	0	0	0	0	2	0	0	0	0	0	0	0	2	0	0	0	0	0	0	0	0	0	1	0	1	0	0	0	0	0
0	0	0	0	0	0	0	0	0	0	1	0	0	1	0	0	0	0	0	0	0	0	0	0	0	0	0	0	4	0	0	0	0	0
0	0	0	0	0	0	0	0	0	0	0	0	0	0	0	0	0	0	0	0	0	0	0	0	0	0	1	0	1	0	0	0	0	0
0	0	0	0	0	0	0	0	0	0	0	0	0	0	0	0	0	0	0	0	0	0	0	0	0	0	0	0	0	0	0	0	0	0
0	0	0	0	0	0	0	0	0	0	0	0	0	0	0	0	0	0	0	0	0	0	0	0	0	0	0	0	0	0	0	0	0	0
0	0	5	1	0	0	1	0	0	0	2	0	0	5	3	3	0	0	0	0	0	0	0	0	0	0	45	0	28	1	0	0	0	0
0	0	1	4	0	0	0	0	0	0	1	0	0	0	0	0	0	0	0	0	0	0	0	0	0	0	2	0	1	0	0	0	0	0
0	0	0	1	0	0	0	0	0	0	1	0	1	0	0	0	0	0	0	0	0	0	0	0	0	0	0	0	3	0	0	0	0	0
0	0	2	0	0	0	0	0	0	0	0	0	0	1	0	0	0	0	0	0	0	0	0	0	0	0	1	0	0	0	0	0	0	0
0	0	0	0	0	0	0	0	0	0	0	0	0	0	0	0	0	0	0	0	0	0	0	0	0	0	0	0	1	0	0	0	0	0
0	0	0	0	0	0	0	0	0	0	0	0	0	3	0	0	0	0	0	0	0	0	0	0	0	0	14	0	5	0	0	0	0	0
0	0	0	0	0	0	0	0	0	0	0	0	0	0	0	0	0	0	0	0	0	0	0	0	0	0	0	0	1	0	0	0	0	0
0	0	0	1	0	0	0	0	0	0	0	0	0	5	0	0	2	0	0	0	1	0	0	0	0	0	10	0	1	0	0	0	0	0
0	0	0	0	0	0	0	0	0	0	0	0	0	0	0	0	0	0	0	0	0	0	0	0	0	0	0	0	0	0	0	0	0	0
0	0	2	0	0	0	0	0	0	0	8	0	1	5	0	4	11	0	0	0	0	0	0	0	0	0	5	3	0	0	0	0	0	0
0	0	1	2	0	0	0	0	0	0	0	0	0	4	0	1	1	0	0	0	0	0	0	0	0	0	4	0	0	0	0	0	0	0
0	0	0	0	0	0	0	0	0	0	1	0	0	0	0	0	0	0	0	0	0	0	0	0	0	0	0	0	0	0	0	0	0	0
0	0	0	0	0	0	0	0	0	0	0	0	0	19	1	0	0	0	0	0	0	0	0	0	0	0	10	0	11	0	0	0	0	0
0	0	1	0	0	0	0	0	0	0	0	0	0	2	1	0	0	0	0	0	0	0	0	0	0	0	1	0	2	0	0	0	0	0
0	0	0	0	0	0	0	0	0	0	0	0	0	1	0	0	0	0	0	0	0	0	0	0	0	0	0	0	0	0	0	0	0	0
0	0	0	0	0	0	0	0	0	0	0	0	0	0	0	0	0	0	0	0	0	0	0	0	0	0	0	0	0	0	0	0	0	0

Table 4.5 Data matrix used for the cluster analysis with foraminifers (continued).

Env. setting	*Species*	*ST1-0*	*ST1-1*	*ST1-2*	*ST1-3*	*ST1-4*	*ST1-6*	*ST1-7*	*ST2-0*	*ST2-1*	*ST2-2*	*ST2-3*	*ST2-4*	*ST2-5*	*ST3-1*	*ST3-2*	*ST3-3*	*ST3-4*	*ST3-5*	*ST3-6*	*ST3-7*	*ST3-8*	*ST7-0*	*ST7-1*	*ST7-2*	*ST7-5*
M	*Quinqueloculina* aff. *stalkeri*	0	0	0	0	0	0	0	0	0	0	0	0	0	0	3	3	0	0	0	0	0	0	2	0	0
M	*Quinqueloculina berthelotiana*	1	0	0	0	0	1	0	1	1	0	4	5	4	2	3	3	0	1	0	0	0	0	4	0	0
M	*Quinqueloculina lata*	1	0	0	0	0	0	0	0	0	1	0	3	1	0	1	0	0	0	0	0	0	2	1	0	0
M	*Quinqueloculina milletti*	0	0	0	0	0	0	0	0	0	4	0	3	1	0	1	4	0	0	0	0	0	0	1	0	0
M	*Quinqueloculina nodulosa*	0	0	0	0	0	0	0	0	0	0	0	0	0	0	8	0	0	1	0	0	0	0	1	0	0
M	*Quinqueloculina parvula*	1	0	1	0	0	0	0	0	0	0	0	0	0	0	8	3	0	0	0	0	0	3	2	0	0
M	*Quinqueloculina peseudobuchiana*	0	0	0	0	0	0	0	0	0	2	0	2	0	0	0	0	0	0	0	0	0	0	0	0	0
M/L	*Quinqueloculina seminulum*	3	1	0	0	0	0	0	3	0	0	0	9	4	0	3	2	0	0	0	0	0	0	4	0	0
M	*Quinqueloculina* cf. *subpolygona*	1	1	0	0	0	0	0	2	1	5	0	8	13	15	15	21	1	3	0	0	0	0	1	0	0
M	*Quinqueloculina trigonula*	3	2	0	1	0	0	0	1	1	8	7	10	11	17	21	12	0	0	0	0	0	6	17	0	0
M	*Quinqueloculina ungeriana*	0	0	0	0	0	0	0	0	0	0	0	0	0	0	0	0	0	0	0	0	0	0	0	0	0
M	*Quinqueloculina vulgaris*	0	0	0	0	0	0	0	0	0	0	0	0	0	0	0	0	0	0	0	0	0	0	0	0	0
M	*Rectoglandulina* sp.	0	0	0	0	0	0	0	0	0	0	0	0	0	0	0	0	0	0	0	0	0	0	0	0	0
M	*Reussella spinulosa*	0	0	0	0	0	0	0	0	0	1	0	0	0	1	3	0	0	1	0	0	0	2	1	0	0
M	*Rosalina bradyi*	0	1	0	0	0	0	0	1	3	3	0	4	6	12	8	7	0	4	0	0	0	8	15	0	0
M	*Rosalina floridana*	0	0	0	0	0	0	0	3	2	4	1	4	10	13	5	12	0	0	0	0	0	1	7	0	0
M	*Rosalina globularis*	0	0	0	0	0	0	0	0	0	0	0	3	5	0	0	0	0	0	0	0	0	2	4	0	0
M	*Sigmoilinita costata*	0	0	0	2	0	0	0	1	0	3	3	7	9	10	8	9	0	0	0	0	0	5	11	0	0
M	*Sinuloculina consobrina*	0	0	0	0	0	0	0	0	0	1	0	2	1	1	1	3	0	0	0	0	0	0	3	0	0
M	*Sinuloculina cyclostoma*	0	0	1	0	0	0	0	0	0	1	0	1	1	1	5	1	0	0	0	0	0	2	0	0	0
M	*Sinuloculina inflata*	0	0	0	0	0	0	0	0	0	0	0	0	0	1	0	0	0	1	0	0	0	0	4	0	0
M	*Sinuloculina rotunda*	0	0	0	0	0	0	0	0	1	1	0	0	1	1	0	4	0	0	0	0	0	1	0	0	0
M	*Siphonaperta agglutinans*	0	0	0	0	0	0	0	0	0	0	1	0	0	2	1	3	0	0	0	0	0	0	1	0	0
M	*Siphonaperta aspera*	0	0	0	0	0	0	0	0	0	0	0	1	1	2	0	0	0	0	0	0	0	0	1	0	0
M	*Siphonaperta irregularis*	0	0	0	0	0	0	0	0	0	0	1	3	0	0	0	0	0	0	0	0	0	0	1	0	0
M	*Siphonaperta torrei*	0	0	0	0	0	0	0	0	0	0	2	3	5	2	0	0	0	0	0	0	0	2	0	0	0
M	*Siphonina reticulata*	0	0	0	0	0	0	0	0	0	0	0	1	1	2	0	0	0	0	0	0	0	3	3	0	0
M	*Sorites orbiculus*	0	1	0	0	0	0	0	0	1	1	0	2	3	2	4	0	0	0	0	0	0	0	1	0	0
M	*Spiroloculina antillarum*	1	0	1	1	0	0	0	4	0	2	0	2	8	7	3	3	0	0	0	0	0	2	5	0	0
M	*Spiroloculina* cf. *dilatata*	0	0	0	0	0	0	0	1	0	0	0	0	6	5	0	16	0	0	0	0	0	0	1	0	0
M	*Spiroloculina excavata*	0	0	0	0	0	0	0	1	0	0	0	0	2	1	15	0	0	1	0	0	0	0	0	0	0
M	*Spiroloculina ornata tricarinata*	0	0	0	0	0	0	0	0	0	1	0	2	1	0	0	2	0	0	0	0	0	0	3	0	0
M	*Textularia bocki*	0	0	0	0	0	0	0	1	1	0	0	0	1	2	1	0	0	0	0	0	0	0	0	0	0
M	*Textularia calva*	0	0	0	0	0	0	0	1	0	0	0	0	2	1	0	0	0	0	0	0	0	0	1	0	0
M	*Textularia conica*	0	0	0	0	0	0	0	2	0	0	2	2	0	7	2	2	0	0	0	0	0	2	1	0	0
M	*Textularia* sp.1	0	0	0	0	0	0	0	1	0	2	0	0	2	2	0	1	0	0	0	0	0	0	0	0	0
M	*Triloculina marioni*	4	2	0	0	1	0	0	11	3	6	7	13	10	17	13	11	0	4	0	0	0	9	19	0	0
M	*Triloculina plicata*	0	1	0	0	0	0	0	1	2	1	4	2	2	2	1	2	0	0	0	0	0	2	2	0	0
M	*Triloculina tricarinata*	0	0	0	0	0	0	0	1	0	0	1	1	2	2	1	2	0	0	0	0	0	0	1	0	0
M/L	*Triloculina trigonula*	0	0	0	0	0	0	0	7	2	9	3	5	3	7	7	7	0	2	0	0	0	1	7	0	0
M	*Uvigerina* sp.	0	0	0	0	0	0	0	1	0	0	0	3	6	7	2	0	0	0	0	0	0	2	8	0	0
M	*Vertebralina striata*	0	0	0	0	0	0	0	1	0	1	2	1	9	10	16	11	1	2	0	0	0	0	0	0	0
M	*Wiesnerella auriculata*	0	0	0	0	0	0	0	1	0	0	1	2	0	2	0	0	0	0	0	0	0	1	2	0	0

1981; Carbonel and Pujos 1982; Ippoliti 1993–4; Slack *et al.* 1995). In addition to the above, studies of the following provide database sources of marine Mediterranean ostracods: Gulf of Naples (Müller 1894), Adriatic Sea (Bonaduce *et al.* 1975; Breman 1975), Italian seas (Montenegro *et al.* 1998), and Cyprus and Naxos Island, Greece (Alhersuch 1979; Barbeito-Gonzalez 1971). Other useful investigations include those of Nile delta margin core samples that provide information on pre-modern Quaternary ostracods (Sneh *et al.* 1986; Pugliese and Stanley 1991), and Quaternary ostracods of Monastir, Tunisia (Wouters 1973).

The ostracod fauna, found in 60 of the 78 core samples, consist of 36 species. These are listed in alphabetical order in the Table 4.6. Most species of ostracods identified in this study are well known in the modern Mediterranean. Species have been included in one of several well-defined groups: **freshwater, brackish, brackish–infralittoral,**

ST7-6	ST7-8	ST9-1	ST9-2	ST9-3	ST9-4	ST9-5	ST9-6	ST9-7	ST9-8	ST12-0	ST12-1	ST12-1B	ST13-1	ST13-2	ST13-3	ST13-4	ST13-5	ST13-7	ST13-8	ST13-9	ST13-10	ST13-11	ST13-12	ST13-14	ST13-15	ST14-1	ST14-2	ST17-1	ST17-2	ST17-3	ST17-4	ST17-5	ST17-6
0	0	1	0	0	0	0	0	0	0	0	0	0	0	0	0	0	0	0	0	0	0	0	0	0	0	0	0	0	0	0	0	0	0
0	0	0	0	0	0	0	0	0	0	0	0	0	1	0	1	0	0	0	0	0	0	0	0	0	0	4	0	2	0	0	0	0	0
0	0	0	0	0	0	0	0	0	0	0	0	0	0	0	0	0	0	0	0	0	0	0	0	0	0	0	0	2	0	0	0	0	0
0	0	3	0	0	0	0	0	0	0	2	0	0	3	0	0	0	0	0	0	0	0	0	0	0	0	0	0	2	1	1	3	0	0
0	0	0	0	0	0	0	0	0	0	0	0	0	0	0	0	0	0	0	0	0	0	0	0	0	0	2	0	1	0	0	0	0	0
0	0	0	0	0	0	0	0	0	0	0	0	0	0	0	0	0	0	0	0	0	0	0	0	0	0	4	0	0	0	0	0	0	0
0	0	0	0	0	0	0	0	0	0	0	0	0	0	0	0	0	0	0	0	0	0	0	0	0	0	0	0	0	0	0	0	0	0
0	1	1	0	0	1	0	1	2	0	2	0	0	2	0	0	0	0	0	0	0	0	0	0	0	0	4	0	2	0	0	0	0	0
0	0	0	1	0	0	0	0	0	0	3	0	0	3	0	0	0	0	0	0	0	0	0	0	0	0	5	0	0	0	0	0	0	0
0	0	1	0	0	0	0	0	0	0	0	0	0	6	0	0	4	0	0	0	0	0	0	0	0	0	22	0	16	0	1	0	0	0
0	0	1	5	0	0	0	0	0	0	3	0	2	4	0	0	0	0	0	0	0	0	0	0	0	0	0	0	0	0	0	0	0	0
0	0	3	4	0	0	0	0	0	0	0	0	0	0	0	0	0	0	0	0	0	0	0	0	0	0	3	0	1	0	0	0	0	0
0	0	0	0	0	0	1	0	0	2	0	0	0	0	0	0	0	0	0	0	0	0	0	0	0	0	0	0	0	0	0	0	0	0
0	0	0	0	0	0	0	0	0	0	0	0	0	0	0	0	0	0	0	0	0	0	0	0	0	0	0	0	0	0	0	0	0	0
0	0	0	1	0	0	0	0	0	8	0	0	1	8	0	1	2	0	3	0	0	0	0	0	0	1	10	0	4	0	0	0	0	0
0	0	0	1	0	0	0	0	0	0	1	0	0	5	2	0	0	0	0	0	0	0	0	0	0	0	3	0	4	0	0	0	0	0
0	0	0	0	0	0	0	0	0	0	0	0	0	0	0	0	0	0	0	0	0	0	0	0	0	0	2	0	1	0	0	0	0	0
0	0	0	1	0	0	0	0	0	0	0	0	0	1	0	1	0	0	0	0	0	0	0	0	0	0	5	0	5	0	0	0	0	0
0	0	0	1	0	0	0	0	0	0	0	0	0	0	0	0	0	0	0	0	0	0	0	0	0	0	0	0	0	0	0	0	0	0
0	0	1	0	0	0	0	0	0	0	1	0	1	0	0	0	0	0	0	0	4	0	0	0	0	0	0	0	0	0	1	0	0	0
0	0	0	0	0	0	0	0	0	0	0	0	0	0	0	0	0	0	0	0	0	0	0	0	0	0	0	0	0	0	0	0	0	0
0	0	0	0	0	0	0	0	0	0	1	0	0	0	0	0	0	0	0	0	0	0	0	0	0	0	2	0	0	0	0	0	0	0
0	0	0	0	0	0	0	0	0	0	0	0	0	2	0	0	0	0	0	0	0	0	0	0	0	0	0	0	0	0	0	0	0	0
0	0	0	0	0	0	0	0	1	0	0	0	0	0	0	0	0	0	0	0	0	0	0	0	0	0	0	0	0	0	0	0	0	0
0	0	0	0	0	0	0	0	0	0	0	0	0	0	0	0	0	0	0	0	0	0	0	0	0	0	0	0	0	0	0	0	0	0
0	0	0	0	0	0	0	0	0	0	0	0	0	1	0	0	0	0	0	0	0	0	0	0	0	0	0	0	0	0	0	0	0	0
0	0	0	0	0	0	0	0	0	0	0	0	0	0	0	0	0	0	0	0	0	0	0	0	0	0	0	0	0	0	0	0	0	0
0	0	0	0	0	0	0	0	0	0	0	0	0	0	0	0	0	0	0	0	0	0	0	0	0	0	4	0	2	0	0	0	0	0
0	0	1	0	0	0	0	0	0	0	1	2	0	3	3	1	2	0	0	0	0	0	0	0	0	0	3	0	9	0	0	0	0	0
0	0	0	1	0	0	0	0	0	0	0	0	0	2	0	1	0	0	0	0	0	0	0	0	0	0	0	0	2	0	0	0	0	0
0	0	0	0	0	0	0	0	0	0	0	0	0	0	0	0	0	0	0	0	0	0	0	0	0	0	0	0	0	0	0	0	0	0
0	0	0	0	0	0	0	0	0	0	0	0	0	0	0	0	0	0	0	0	0	0	0	0	0	0	1	0	0	0	0	0	0	0
0	0	0	0	0	0	0	0	0	0	0	0	0	3	0	0	0	0	0	0	0	0	0	0	0	0	0	0	0	0	0	0	0	0
0	0	0	0	0	0	0	0	0	0	0	0	0	0	0	0	0	0	0	0	0	0	0	0	0	0	0	0	0	0	0	0	0	0
0	0	0	0	0	0	0	0	0	0	0	0	0	4	0	0	0	0	0	0	0	0	0	0	0	0	0	0	0	0	0	0	0	0
0	0	0	0	0	0	0	0	0	0	0	0	0	2	0	0	0	0	0	0	0	0	0	0	0	0	2	0	0	0	0	0	0	0
0	0	0	3	1	0	0	0	0	0	2	1	1	12	1	0	1	0	0	0	1	0	1	0	0	0	17	0	9	0	0	0	0	0
0	0	0	3	1	0	0	0	0	0	0	0	0	4	2	1	0	0	0	0	0	0	0	0	0	0	4	0	2	0	0	0	0	0
0	0	0	1	0	0	0	0	0	0	0	0	0	4	1	0	0	0	0	0	0	0	0	0	0	0	0	0	6	0	0	0	0	0
0	0	3	6	0	0	0	0	0	0	1	2	0	3	0	0	1	0	0	0	0	0	0	0	0	0	12	0	8	0	0	0	0	0
0	0	0	0	0	0	0	0	0	0	0	0	0	2	0	0	0	0	0	0	0	0	0	0	0	0	3	0	3	0	0	0	0	0
0	0	2	0	0	0	0	0	0	0	1	0	0	0	2	0	0	0	0	0	0	0	0	0	0	0	1	0	2	0	0	0	0	0
0	0	0	0	0	0	0	0	0	0	0	0	0	0	0	0	0	0	0	0	0	0	0	0	0	0	0	0	0	0	0	0	0	0

infralittoral and **infralittoral–circalittoral**. Only one species (*Kroemmelbeinia coae*) is not well known, but it has been described in the Upper Pliocene of Kos Island (Mostafavi 1983); rare juveniles were also found by Ippoliti (1993–4) in modern Aboukir Bay. Its similarity to *Microcytherura* suggests that this form can be included in the infralittoral–circalittoral group. The species in open nomenclature (sp., cf.) show morphological features that enable us to compare them to similar modern species.

To interpret palaeoenvironments with ostracods correctly, it is necessary to distinguish autochthonous (lived in place) from allochthonous (displaced) specimens. Following Pugliese and Stanley (1991), a species is autochthonous if it is represented by both adult and young specimens in the same sample. This indicates that most life cycle stages have been preserved *in situ*. However, ostracod autochthony may also be recognised in species represented only by adults but, for this, autochthony needs to be

Table 4.6 Ostracod species identified in this study, with taxa listed in alphabetical order.

Aglaiocypris rara (G.W. Müller, 1894)
Aurila arborescens (Brady, 1865)
Aurila convexa (Baird, 1850)
Aurila fallax Ruggieri, 1984
Aurila cf. *convexa* (Baird, 1850)
Bairdia corpulenta G.W. Müller, 1894
Bairdia longevaginata G.W. Müller, 1894
Basslerites berchoni (Brady, 1870)
Caudites calceolatus (Costa, 1853)
Cistacythereis rubra (G.W. Müller, 1894)
Cyprideis torosa (Jones, 1850)
Cytheretta adriatica Ruggieri, 1952
Darwinula stevensoni (Brady & Robertson, 1870)
Heterocythereis voraginosa Athersuch, 1979
Ilyocypris sp.1
Kroemmelbeinia coae Mostafavi, 1983
Limnocytheridae
Loxoconcha affinis (Brady, 1866)
Loxoconcha elliptica Brady, 1868
Loxoconcha ovulata Costa, 1853
Loxoconcha rhomboidea (Fischer, 1855)
Microcytherura fulva (Brady & Robertson, 1874)
Pontocythere turbida (G.W. Müller, 1894)
Potamocypris sp.
Procytherideis complicata (Ruggieri, 1953)
Procytherides subspiralis (Brady, Crosskey & Robertson, 1874)
Semicytherura cf. *S. sulcata* (G.W. Müller, 1894)
Tenedocythere prava Baird, 1850
Urocythereis britannica Athersuch, 1977
Urocythereis distinguenda Athersuch, 1979
Urocythereis favosa (Roemer, 1838)
Urocythereis cf. *U. britannica* Athersuch, 1977
Xestoleberis communis G.W. Müller, 1894
Xestoleberis gr. *dispar* G.W. Müller, 1894
Xestoleberis fuscomaculata G.W. Müller, 1894
Xestoleberis sp. 2

Table 4.7 Data matrix used for the cluster analysis with ostracods.

Env. setting	*Species*	*ST1-0*	*ST1-1*	*ST1-2*	*ST1-3*	*ST1-4*	*ST1-5*	*ST1-6*	*ST1-7*	*ST2-0*	*ST2-4*	*ST2-5*	*ST3-1*	*ST3-2*	*ST3-3*	*ST3-4*	*ST3-5*	*ST3-6*	*ST3-7*	*ST3-8*	*ST7-0*	*ST7-1*	*ST7-3*	*ST9-1*	*ST9-2*	*ST9-3*
M	*Aglayocypris rara*	0	0	0	0	0	0	0	0	0	0	0	0	0	0	0	0	0	0	0	0	0	0	0	0	0
L-M	*Aurila arborescens*	0	0	0	0	0	0	0	0	0	0	1	0	0	0	0	0	0	0	0	0	0	0	0	0	0
M	*Aurila* cf. *convexa*	0	0	0	0	0	0	0	0	0	0	0	0	0	2	0	0	0	0	0	0	0	0	0	0	0
M	*Aurila convexa*	0	0	0	0	0	0	0	0	0	2	1	0	0	0	0	0	0	0	0	0	0	0	0	0	0
M	*Aurila fallax*	0	0	0	0	0	0	0	0	2	0	0	0	1	0	0	0	0	0	0	0	0	0	0	0	0
M	*Bairdia corpulenta*	0	0	0	0	0	0	0	0	0	0	0	0	0	1	0	0	0	0	0	0	0	0	0	0	0
M	*Bairdia longevaginata*	0	0	0	0	0	0	0	0	0	0	0	1	1	0	0	0	0	0	0	0	0	0	0	1	0
L-M	*Baslerites berchoni*	0	0	0	0	0	0	0	0	0	0	0	0	0	0	0	0	0	0	0	0	0	0	0	0	0
M	*Caudites calceolatus*	0	0	0	0	0	0	0	0	0	0	0	0	0	0	0	0	0	0	0	0	1	0	0	0	0
M	*Cistacythereis rubra*	0	0	0	0	0	0	0	0	0	0	0	0	0	1	0	0	0	0	0	0	0	0	0	0	0
L	*Cyprideis torosa*	0	0	2	0	0	1	1	2	0	0	0	0	0	11	0	157	7	6	5	0	0	1	0	0	85
M	*Cytheretta adriatica*	0	0	0	0	0	0	0	0	0	1	4	1	0	3	0	0	0	0	0	1	0	0	0	0	0
FW	*Darwinula stevensoni*	0	0	0	0	0	0	0	0	0	0	0	0	0	1	0	0	0	0	0	0	0	0	0	0	0
M	*Heterocythereis voraginosa*	0	0	0	0	0	0	0	0	0	0	0	0	0	1	0	0	0	0	0	0	0	0	0	1	0
FW	*Ilyocypris* sp.1	0	0	0	0	0	0	0	0	0	0	0	0	0	0	0	0	0	0	0	0	0	0	0	1	0
M	*Kroemmelbeinia coae*	0	0	0	0	0	0	0	0	0	0	1	1	0	0	0	0	0	0	0	0	0	0	0	0	0
FW	*Limnocytheridae*	0	0	0	0	0	0	0	0	0	0	0	0	0	0	0	0	0	0	0	0	0	0	0	0	0
M	*Loxoconcha affinis*	0	0	0	0	0	0	0	0	0	0	13	0	0	6	0	0	0	0	0	0	0	0	0	0	0
L	*Loxoconcha elliptica*	0	0	0	0	0	0	0	0	0	0	0	0	0	0	0	0	0	0	0	0	0	0	0	0	0
M	*Loxoconcha ovulata*	0	0	0	0	0	0	0	0	0	0	0	0	0	0	0	0	0	0	0	0	0	0	0	0	0
M	*Loxoconcha rhomboidea*	1	0	0	0	0	0	0	0	0	0	0	0	0	0	0	0	0	0	0	0	0	0	0	0	0
M	*Microcytherura fulva*	0	0	0	0	0	0	0	0	0	0	0	0	0	0	0	0	0	0	0	0	0	0	0	0	0
M	*Pontocythere turbida*	0	0	0	0	0	0	0	0	0	0	15	0	1	0	0	0	0	0	0	0	2	0	0	0	0
M	*Procytherideis complicata*	0	0	0	0	1	0	0	0	0	0	0	0	0	0	0	0	0	0	0	0	0	0	0	0	0
M	*Semicytherura* cf. *sulcata*	0	0	0	0	0	0	0	0	0	0	0	0	0	0	0	0	0	0	0	0	1	0	0	0	0
M	*Tenedocythere prava*	0	0	0	0	0	0	0	0	0	0	2	0	1	0	0	0	0	0	0	0	0	0	0	0	0
M	*Urocythereis britannica*	0	2	0	1	0	0	0	0	0	10	65	0	0	0	0	0	0	0	0	1	0	0	1	2	0
M	*Urocythereis* cf. *britannica*	2	0	0	0	0	0	0	0	3	0	0	22	11	8	1	0	0	0	0	0	0	0	0	0	0
M	*Urocythereis distinguenda*	0	0	0	0	0	0	0	0	0	0	0	0	0	0	0	0	0	0	0	0	0	0	0	0	0
M	*Urocythereis favosa*	0	0	0	0	0	0	0	0	0	0	1	1	0	2	0	0	0	0	0	0	0	0	0	0	0
M	*Xestoleberis communis*	1	0	0	0	0	0	0	0	0	2	11	3	6	6	1	0	0	0	0	1	1	0	2	1	0
M	*Xestoleberis gr. dispar*	0	0	0	0	0	0	0	0	0	0	0	0	0	0	0	0	0	0	0	0	1	0	0	2	0
M	*Xestoleberis* sp.2	0	0	0	0	0	0	0	0	0	0	1	0	0	0	0	0	0	0	0	0	0	0	0	0	0

supported by good carapace preservation, presence of complete carapaces and occurrence of both sexes. In addition, iron-stained ostracods have been found in some samples (see Discussion section). When these oxidised valves occur in a sample together with well-preserved taxa, they are considered allochthonous. In contrast, when oxidised carapaces occur in samples where all the specimens are stained, they are considered autochthonous.

The data matrix used to perform the cluster analysis was compiled taking into account only the autochthonous specimens (Table 4.7).

3 Observations

3.1 Molluscs

Only 46 (63%) of the 73 samples from the nine cores provide identifiable molluscs. The molluscan fauna in the 46 samples contain 53 bivalve and 63 gastropod species, and one species (*Dentalium dentalis*) of scaphopod (Table 4.2). Analysis reveals a total of 87 marine, 6 lagoonal and 5 freshwater species (Table 4.3).

Two major groups of samples are identified on the basis of cluster analysis (Figure 4.4A). **Cluster 1** includes samples characterised by a dominance of marine taxa, such as *Bittium reticulatum*, *Chamelea gallina*, *Glycymeris glycymeris*, *Loripes lacteus*, *Musculus costulatus* and *Tricolia speciosa*. **Cluster 2** comprises samples with taxa of lagoonal environment affinity, and two subclusters are recorded. **Subcluster 2a** is characterised by epifaunal lagoonal species that require vegetate bottoms: these are *Hydrobia stagnalis* and *Potamides conicus*. **Subcluster 2b** includes samples dominated by the infaunal species *Abra ovata* and *Cerastoderma glaucum*. It is also noted that three samples of cluster 2 (ST9-7, ST9-8, ST13-9) denote a freshwater influence, as indicated by the presence of *Corbicula fluminalis*,

ST9-4	*ST9-5*	*ST9-6*	*ST9-7*	*ST9-8*	*ST12-0*	*ST12-1*	*ST12-1B*	*ST12-3*	*ST13-1*	*ST13-2*	*ST13-3*	*ST13-4*	*ST13-5*	*ST13-7*	*ST13-8*	*ST13-9*	*ST13-10*	*ST13-11*	*ST13-12*	*ST13-14*	*ST13-15*	*ST14-1*	*ST14-2*	*ST14-2B*	*ST14-4*	*ST14-5*	*ST17-1*	*ST17-2*	*ST17-3*	*ST17-4*	*ST17-5*	*ST17-6*	*ST17-7*
0	0	0	0	0	0	0	0	0	0	0	0	0	0	0	0	0	0	0	0	0	0	0	0	0	0	0	0	0	1	0	0	0	0
0	0	2	0	0	0	0	0	0	0	3	0	0	2	0	0	0	0	0	0	0	0	0	0	0	0	0	0	0	40	2	0	0	0
0	0	0	0	0	0	0	0	0	0	0	0	0	0	0	0	0	0	0	0	0	0	1	0	0	0	0	4	0	0	0	0	0	0
0	0	0	0	0	0	0	0	0	0	0	0	0	0	0	0	0	0	0	0	0	0	0	0	0	0	0	0	0	0	0	0	0	0
0	0	0	0	0	0	0	0	0	0	0	0	0	0	0	0	0	0	0	0	0	0	0	0	0	0	0	6	0	0	0	0	0	0
0	0	0	0	0	0	0	0	0	0	1	0	0	0	0	0	0	0	0	0	0	0	0	0	0	0	0	0	0	0	0	0	0	0
0	0	0	0	0	0	0	0	0	0	0	0	0	0	0	0	0	0	0	0	0	0	0	0	0	0	0	0	0	0	0	0	0	0
0	0	0	0	0	0	0	0	0	0	1	0	0	0	0	0	0	0	0	0	0	0	0	0	0	0	0	0	0	0	0	0	0	0
0	0	0	0	0	0	0	0	0	0	0	0	0	0	0	0	0	0	0	0	0	0	0	0	0	0	0	0	0	0	0	0	0	0
0	0	0	0	0	0	0	0	0	0	0	0	0	0	0	0	0	0	0	0	0	0	0	0	0	0	0	0	0	0	0	0	0	0
8	55	0	0	46	0	5	19	3	4	0	8	0	29	11	59	0	62	11	1	6	139	24	9	0	1	1	84	155	235	220	31	5	1
0	0	0	0	0	0	0	0	0	0	0	0	0	0	0	0	0	0	0	0	0	0	1	0	0	0	0	0	0	0	0	0	0	0
0	1	0	0	0	0	0	0	0	0	0	0	0	0	0	0	1	0	0	0	0	0	0	0	0	0	0	0	0	0	0	0	0	0
0	0	0	0	0	0	0	0	0	0	0	0	0	0	0	0	0	0	0	0	0	0	0	0	0	0	0	0	0	0	0	0	0	0
0	1	1	0	0	0	0	0	0	0	0	0	0	0	0	0	0	0	0	0	0	0	0	0	0	0	0	0	0	0	0	0	0	0
0	0	0	0	0	0	0	0	0	0	0	0	0	0	0	0	0	0	0	0	0	0	0	0	0	0	0	0	0	0	0	0	0	0
0	1	0	0	0	0	0	0	0	0	0	0	0	0	0	0	0	0	0	0	0	0	0	0	0	0	0	0	0	0	0	0	0	0
0	0	0	0	0	0	0	0	0	3	2	0	0	0	0	0	0	0	0	0	0	0	0	0	0	0	0	0	0	0	0	0	0	0
0	0	0	0	0	0	0	0	0	0	0	0	0	0	0	0	0	0	0	0	0	0	0	0	0	0	0	0	0	0	1	0	0	0
0	0	0	0	0	1	0	0	0	0	0	0	0	0	0	0	0	0	0	0	0	0	0	0	0	0	0	0	0	0	0	0	0	0
0	0	0	0	0	0	0	0	0	0	0	0	0	0	0	0	0	0	0	0	0	0	0	0	0	0	0	0	0	0	0	0	0	0
0	0	0	0	0	0	0	0	0	1	0	0	0	0	0	0	0	0	0	0	0	0	0	0	0	0	0	0	0	0	0	0	0	0
0	0	0	0	0	0	0	0	0	0	0	0	0	0	0	0	0	0	0	0	0	0	1	0	0	0	0	0	0	0	0	0	0	0
0	0	0	0	0	0	0	0	0	0	0	0	0	0	0	0	0	0	0	0	0	0	0	0	0	0	0	0	0	0	0	0	0	0
0	0	0	1	0	0	0	0	0	0	0	0	0	0	0	0	1	0	0	0	0	0	0	0	0	0	0	0	0	0	0	0	0	0
0	0	0	0	0	0	0	0	0	1	0	0	0	0	0	0	0	0	0	0	0	0	0	0	0	0	0	0	0	0	0	0	0	0
0	0	0	1	0	0	0	0	0	9	3	0	1	0	0	0	0	0	0	0	0	0	2	0	0	0	0	4	0	0	0	0	0	0
0	0	0	0	0	0	0	0	0	1	1	0	0	0	0	0	0	0	0	0	0	0	5	0	0	0	0	0	0	0	0	0	0	0
0	0	0	0	0	0	0	0	0	0	0	0	0	0	0	0	0	0	0	0	0	0	0	0	0	0	0	5	0	0	0	0	0	0
0	0	0	0	0	0	0	0	0	0	0	0	0	0	0	0	0	0	0	0	0	0	0	0	1	0	0	1	0	0	0	0	0	0
0	0	0	0	0	0	0	0	0	1	0	0	0	0	0	0	0	0	0	0	0	0	0	0	0	0	0	0	0	0	0	0	0	0
0	0	0	0	0	0	0	0	0	0	0	0	0	0	0	0	0	0	0	0	0	0	0	0	0	0	0	0	0	0	0	0	0	0
0	0	0	0	0	0	0	0	0	0	0	0	0	0	0	0	0	0	0	0	0	0	0	0	0	0	0	0	0	0	0	0	0	0

Melanoides tuberculata and *Teodoxus niloticus*. Characteristic taxa are shown in Figure 4.5.

It is of note that 11 core samples contain only poorly preserved molluscan shell debris, commonly fragmented valves with decalcified and/or weathered surfaces and a light grey to off-white colour. These occur at depths of >1.5 m from core tops. With the exception of those in core ST7, worn fragments usually occur in dark, organic-rich mud sections of cores that are, for the most part, identified as lagoonal environment. Moreover, 14 samples are devoid of any molluscan remains, primarily in the mid and lower parts of organic-rich mud core sections. It appears that most of these fauna-poor and mollusc-barren samples, including those without any residue, occur in sections that are dated as older than 3000 yrs BP.

3.2 Foraminifers

The foraminiferal species observed are characteristic mainly of brackish and infralittoral settings. Among these are some species indicative of specific well-defined environments in Mediterranean areas, and on the basis of this environmental information it is possible to define the following groups:

- **Brackish water group** (= L of molluscs) comprises *Ammonia tepida* either exclusively, or, alternatively, as the dominant species in assemblages that tolerate low salinity and are commonly found in shallow marine, lagoon, tidal flat and deltaic Mediterranean settings.

- **Brackish water–infralittoral group** (= L-M of molluscs) includes *Ammonia tepida, A. papillosa* and *A. parkinsoniana* dominant, *Buccella granulata, Cycloforina costata, Elphidium excavatum, E. granosum, Quinqueloculina [?] seminulum* and *Triloculina trigonula*, subordinate.

- **Infralittoral group** (= M of molluscs) comprises species which are normally found in shallow marine settings; some of them are generally related to bottoms characterised by phytophylous algae and marine fanerogams (*). The species are: *Adelosina carinata-striata, A. cliarensis* (*), *A. dubia* (*), *Affinetrina bermudezi, Ammonia beccarii, Amphistegina madascariensis, Buccella granulata* (*), *Cibicidella variabilis* (*), *Cibicides lobatulus* (*), *Cycloforina costata* (*), *C. schlumbergeri, Cymbaloporetta brady, Elphidium* spp. (*), *Glabratella* spp., *Heterostegina antillarum, Lachlanella variolata* (*), *Massilina gualtieriana* (*), *M. secans* (*), *Miliolinella dilatata, M. webbiana* (*), *Peneroplis pertusus* (*), *P. planatus* (*), *Planorbulina mediterranensis* (*), *Rosalina* spp. (*), *Sinuloculina* spp. (*),*Quinqueloculina [?] ungeriana, Schackoinella imperatoria, Siphonaperta aspera* (*), *Sorites orbiculus, Spiroloculina antillarum* (*), *Triloculina plicata* (*), *T. trigonula* (*) and *Vertebralina striata* (*).

- **Infralittoral–circalittoral group** (= M of molluscs) includes species that are characteristic of a large depth range distribution, although some of these are generally distrib-

Figure 4.5 Some molluscan species representative of the Holocene environments recorded in the core samples: (1) *Abra ovata*, external view of the left valve, scale bar = 2 mm; (2) *A. ovata*, internal view of the right valve, bar = 500 mm; (3) *Cerastoderma glaucum*, external view of the right valve, bar = 200 mm; (4) *Hydrobia stagnalis*, scale bar = 500 mm; (5) *H. stagnalis*, scale bar = 500 mm; (6) *Potamides conicus*, protoconch, scale bar = 2 mm; (7) *P. conicus*, scale bar = 100 mm; (8) *Chamelea gallina*, external view of the left valve, scale bar = 500 mm; (9) *C. gallina*, internal view of the left valve, scale bar = 500 mm; (10) *Bittium reticulatum*, scale bar = 2 mm; (11) *Loripes lacteus*, external view of the right valve, scale bar = 500 mm; (12) *L. lacteus*, internal view of the left valve, scale bar = 1 mm; (13) *Corbicula fluminalis*, internal view of the right valve, scale bar = 1 mm.

uted in an upper circalittoral setting (°). The species are: *Adelosina longirostra*, *A. mediterranensis* (°), *A. pulchella*, *Affinetrina alcidi* (°), *Articulina mucronata* (°), *Asterigerinata mamilla* (°), *Astrononion stelligerum*, *Bolivina* spp., *Bulimia elongata*, *Cycloforina tenuicollis* (°), *Discorbinella bertheloti*, *Elphidium macellum*, *E. punctatum*, *E. granosum*, *Planoglabratella opercularis*, *Sinuloculina inflata* (°), *S. rotonda*, *Spiroloculina canaliculata*, *S. excavata*, *Textularia calva*, *T. conica* and *Triloculina trigonula* (°).

Some representative taxa of the Holocene environments recovered in the cores are represented in Figures 4.6 and 4.7.

Cluster analysis serves to identify two major groups (Figure 4.4B).

- **Cluster 1** includes core samples in which the assemblage is highly diversified and the number of specimens is abundant. The foraminifers are mainly represented by

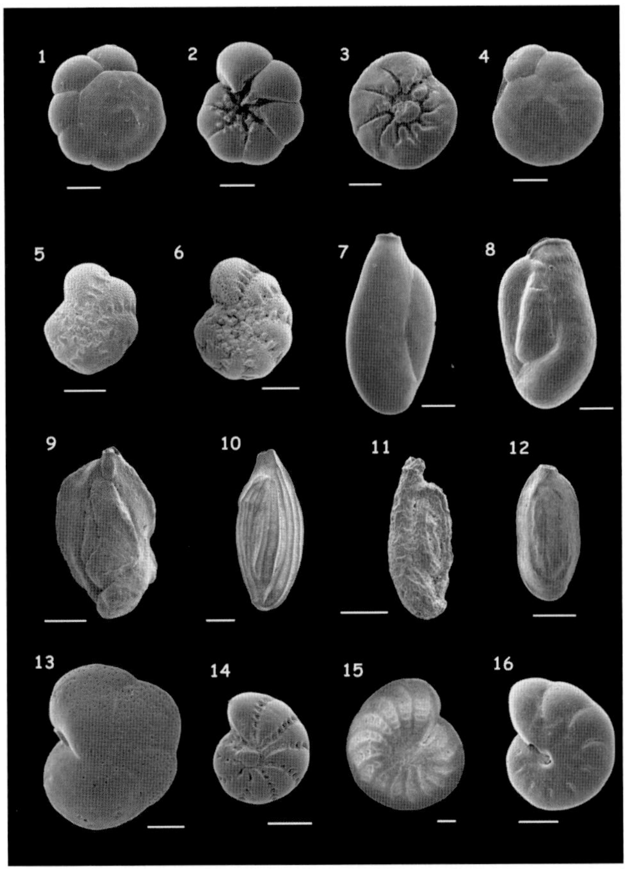

Figure 4.6 Some foraminifer species representative of Holocene environments recorded in the core samples (scale bar = 100 μm): (1) *Ammonia tepida*, spiral side; (2) *A. tepida*, umbilical side; (3) *Ammonia parkinsoniana*, umbilical side; (4) *A. parkinsoniana*, spiral side; (5) *Elphidium granosum*, side view; (6) *Elphidium excavatum*, side view; (7) *Adelosina dubia*, chamber side; (8) *Affinetrina bermudezi*, chamber side; (9) *Cycloforina* aff. *juleana*, chamber side; (10) *Cycloforina costata*, chamber side; (11) *Cycloforina tenuicollis*, chamber side; (12) *Cycloforina schlumbergeri*, chamber side; (13) *Cibicides lobatulus*, umbilical side; (14) *Elphidium gerthi*, side view; (15) *Elphidium jenseni*, side view; (16) *Astrononion stelligerum*, side view.

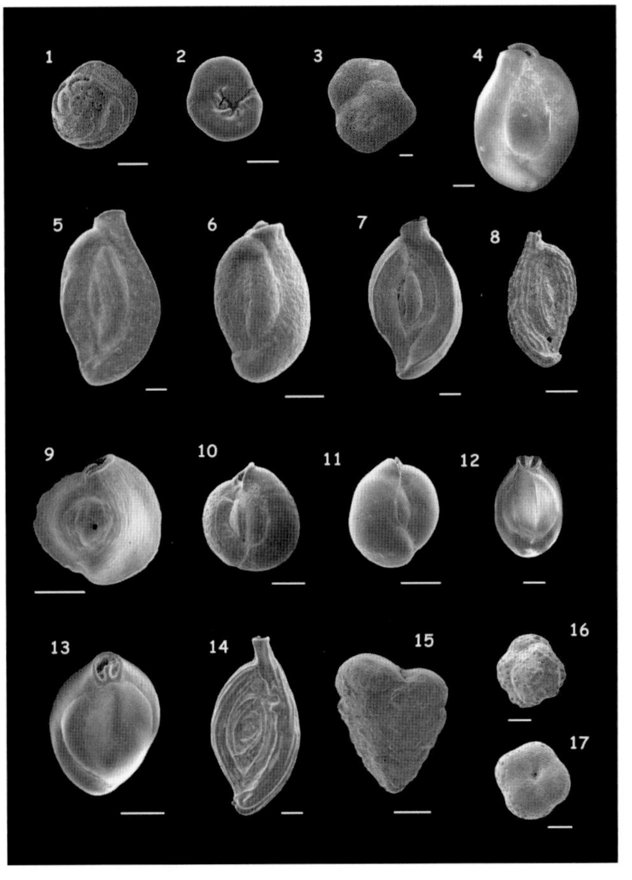

Figure 4.7 Some foraminifer species representative of Holocene environments recorded in core samples (scale bar = 100 μm): (1) *Rosalina bradyi*, spiral side; (2) *R. bradyi*, umbilical side; (3) *Rosalina floridana*, umbilical side; (4) *Massilina gualtieriana*, side view; (5) *Quinqueloculina* cf. *subpolygona*, side view; (6) *Quinqueloculina parvula*, side view; (7) *Spiroloculina* cf. *dilatata*, side view; (8) *Sigmoilina costata*, side view; (9) *Quinqueloculina trigonula*, side view; (10) *Miliolinella dilatata*, side view; (11) *Miliolinella subrotunda*, side view; (12) *Triloculina marioni*, front view; (13) *Triloculina trigonula*, front view; (14) *Spiroloculina antillarum*, side view; (15) *Textularia bocki*, side view; (16) Genus A, spiral side; (17) Genus A, umbilical side.

Ammonia tepida, which is almost always present together with *A. parkinsoniana*, *Peneroplis* spp. and Miliolidae (*Quinqueloculina* spp., *Miliolinella* spp., *Triloculina* spp.). *Massilina gualtieriana* is abundant in some samples, especially in cores ST2 and ST3. On the basis of *A. tepida* abundance, it is possible to distinguish two subclusters (1a and 1b). **Subcluster 1a** is defined by a larger proportion of *A. tepida* tests, together with lower number of species than subcluster 1b. This subcluster indicates a brackish water influence in proximity to a marine setting. **Subcluster 1b**, in contrast, records distinct marine conditions.

- **Cluster 2** is characterised by the dominance of *Ammonia tepida* with *A. parkinsoniana*, *Elphidium excavatum* and scattered occurrence of Miliolidae. It is possible to divide this cluster into two subclusters (2a and 2b). In **subcluster 2a,** three environmental possibilities with lagoonal taxa are recognised: one in which taxa in most samples indicate a clearly brackish (lagoonal) setting, either with *A. tepida* exclusively or as a dominant species; a second, in some other core samples, is defined by the occurrence of *Elphidium* species (infralittoral taxa), thus indicating a marine influence; and a third, somewhat more marine setting, in which other marine species occur, such as *Quinqueloculina* spp. and *Rosalina* spp. **Subcluster 2b** corresponds to samples characterised by few species and specimens. In these samples, a miscellaneous assemblage of species occurs, and it is not possible to assign a single ecological interpretation (one that is either specifically lagoonal or marine).

In sum, Cluster 1a indicates marine conditions with lagoonal influence (= M/L of molluscs), while cluster 1b clearly records marine settings (= M of molluscs). Cluster 2, with three subgroupings, indicates lagoonal environments (= L of molluscs); these three subgroups record increasing marine influences (= L/M of molluscs). Letter symbols are shown in Figures 4.2 and 4.3.

3.3 Ostracods

As has been noted in earlier studies (Pugliese and Stanley 1991), some ostracod groups can be reliably related to their life habitat. For the ostracod data to be compared with foraminiferal and molluscan fauna, it is important to combine the three by using the environmental/benthic bionomy terminology of Pérès and Picard (1964). Ostracod groupings used here also take into account earlier studies (Pugliese and Stanley 1991; Montenegro and Pugliese 1996; Montenegro *et al.* 1998). For the present study, most species found in core samples can be placed into one of five groups.

- **Freshwater group** (= FW of molluscs) consists of the following species: *Darwinula stevensoni*, *Ilyocypris* sp.1, Limnocytheridae and *Potamocypris* sp.

- **Lagoonal group** (= L of molluscs) is constituted by *Cyprideis torosa* and *Loxoconcha elliptica*

- **Lagoonal–infralittoral group** (= L-M of molluscs) includes species which are able to live in both brackish water and marine coastal settings that are freshwater influenced, including *Aurila arborescens* and *Basslerites berchoni*.

- **Infralittoral group** (= M of molluscs) comprises species that generally live on bottoms characterised by phytophylous algae and marine fanerogams. Species denoted by asterisk normally prefer bottoms without vegetation. Most species tolerate nearby inflows of fresh or brackish water. The species include: *Aglaiocypris rara*, *Bairdia longevaginata*, *Caudites calceolatus*, *Cytheretta adriatica*, *Heterocythereis voraginosa*, *Loxoconcha affinis*, *Pontocythere turbida**, *Procytherideis complicata*, *P. subspiralis*, *Semicytherura* cf. *sulcata*, *Tenedocythere prava**, *Urocythereis britannica**, *U. distinguenda**, *U. favosa**, and *U.* cf. *britannica**.

- **Infralittoral–circalittoral group** (= M of molluscs) includes species characterised by a large depth range on the shelf, and these usually live in the first 100 m. Although distributed over a wide range, most forms in this group are also commonly recorded in shallow marine settings. In this study it is noted that species are not associated with deep-water ostracods, but rather indicate shallow settings, i.e. those which can be vegetated as indicated by normally phytal species (°). The species include: *Aurila convexa*, *A. fallax*, *Bairdia corpulenta*, *Cistacythereis rubra*, *Kroemmelbeinia coae*, *Loxoconcha ovulata*, *L. rhomboidea*, *Microcytherura fulva*, *Xestoleberis communis* (°), *Xestoleberis* gr. *dispar* (°), *Xestoleberis* sp.2 (°).

Some ostracod species representative of different major Holocene environments in the studied core are shown in Figure 4.8.

Cluster analysis serves to identify two major groups (Figure 4.4C): cluster 1 (**brackish water**) and cluster 2 (**marine**).

- **Cluster 1** is characterised by *Cyprideis torosa*, one of the most common brackish water ostracods. This cluster can be subdivided into three subclusters: **Subcluster 1a** is monospecific: *Cyprideis torosa* is represented by few specimens, and probably indicates a brackish water setting (wetland, lagoon, etc.). **Subcluster 1b** includes abundant to very abundant specimens of *C. torosa* which are sometimes associated with infralittoral (sample ST9-3) or brackish water–infralittoral (samples ST17-3 and ST17-4) species. These indicate brackish water or other deltaic-related wetland settings that are influenced by the sea. In addition, sample ST9-5 may record a freshwater influence as indicated by the occurrence of freshwater species associated with *C. torosa*. **Subcluster 1c** denotes a well-diversified ostracod fauna, and its assemblage is characterised by *C. torosa* and the marine forms *Urocythereis britannica* and *Aurila* cf. *convexa*. These are associated with several infralittoral and infralittoral–circalittoral species, indicating deltaic-related settings near the sea. The rare freshwater specimens occurring in core sample ST3-3 suggest a nearby freshwater influence.

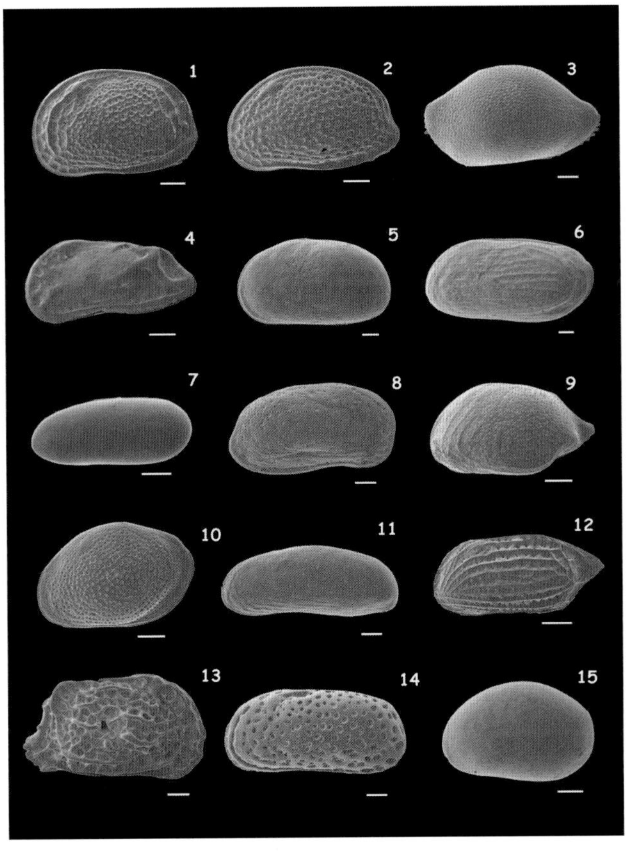

Figure 4.8 Some ostracod species representative of different Holocene environments in studied core samples (scale bar = 100 μm): (1) *Aurila arborescens*, left valve in exterior view; (2) *Aurila convexa*, left valve in exterior view; (3) *Bairdia corpulenta*, left valve in exterior view; (4) *Caudites calceolatus*, left valve in exterior view; (5) *Cyprideis torosa*, left valve in exterior view; (6) *Cytheretta adriatica*, left valve in exterior view; (7) *Darwinula stevensoni*, left valve in exterior view; (8) *Heterocythereis voraginosa*, right valve in exterior view; (9) *Kroemmelbeinia coae*, left valve in exterior view; (10) *Loxoconcha affinis*, left valve in exterior view; (11) *Pontocythere turbida*, left valve in exterior view; (12) *Semicytherura* cf. *sulcata*, left valve in exterior view; (13) *Tenedocythere prava*, right valve in exterior view; (14) *Urocythereis britannica*, left valve in exterior view; (15) *Xestoleberis communis*, left valve in exterior view.

- **Cluster 2** is characterised mostly by infralittoral and infralittoral–circalittoral and, subordinately, lagoonal–infralittoral species. Two subclusters are distinguished. **Subcluster 2a** is characterised by *Urocythereis* cf. *britannica* that is associated with various taxa; it indicates shallow marine and, in some instances, vegetated settings by the presence of *Xestoleberis communis*. **Subcluster 2b** is represented mostly by *U. britannica*, except for sample ST7-1. The several marine species associated with these forms indicate infralittoral settings that, in some cases, are affected by nearby brackish water input to explain the presence of lagoonal taxa (samples ST2-5, ST13-1, ST13-2, ST9-7 and ST9-2).

A third subgroup includes five samples (ST14-2B, ST1-4, ST12-0, ST9-6 and ST13-9) that are not clustered; these are represented by one or two rarely present species. The modern life habitat of these forms indicates, tentatively, infralittoral settings for samples ST14-2B, ST1-4 and ST12-0, and lagoonal (sample ST9-6) and infralittoral (sample ST13-9) settings, both of which are freshwater influenced.

In 22 core samples (from four cores at Heracleion, two at East Canopus), some ostracod specimens are partially to highly oxidised. Most of these samples (18) were collected in organic-rich mud, whereas four were from sandy sections (Figures 4.2, 4.3). No close correlation is recorded between the presence of oxidised ostracod specimens and depth in cores or age of sections. However, in a number of cases, such oxidised-rich samples appear at the same depths as those of samples where preservation of molluscs is poor. For example, 9 (40%) of the 22 oxidised ostracod samples are the same that record poorly preserved molluscs.

4 Discussion

Most of the two molluscan and five microfossil (foraminifers, ostracods) biofacies of Holocene age recovered in Aboukir Bay core samples are attributed to generally comparable lagoonal and shallow marine environments. The core sections record a generally restricted water depth range, from just above and/or about sea level to several metres below sea level. The greatest value of the faunal study is that it provides useful data on salinity changes, sometimes marked ones, within the relatively short (<6 m long) available core sections. Most information on changes through time of depositional environments of substrates under the two cities in the study area pertains to salinity. Our survey shows that, in a number of core samples, the up-core trends of the three different fauna are similar with regard to environmental interpretation. This indicates that biofacies data, obtained independently, are coherent and can serve as reliable environmental markers (Figures 4.2–4.4).

4.1 Temporal lagoonal to marine facies change

Most core samples collected from the sediment substrates at both Heracleion and East Canopus comprise mainly lagoonal molluscs, foraminifers and ostracods (Figures 4.2, 4.3). This is not unexpected, since the two settlements were located at Canopic channel mouths where the dominant, most widespread facies in such low-lying Nile delta margin settings, just landward of the shoreline, were primarily lagoonal-marsh wetland (Bernasconi and Stanley 1994; Stanley and Warne 1998). We find fauna with lagoonal affinities, including wetland with freshwater influence, in the lower parts of some core sections; these evolve upward in cores to marine biofacies. The Holocene lagoonal biofacies described in the borings, as those in modern coastal margin deposits landward of the shoreline (cf. Chen *et al.* 1992; Warne and Stanley 1993; Stanley *et al.* 2004), occur primarily in mud-rich core sections of early to late Holocene age. These facies, normally comprising dark-grey and olive coloured, plant-rich clayey silts and silty clays, have also been described in other studies of the Nile delta (Bernasconi *et al.* 1991; Loizeau and Stanley 1993; Bernasconi and Stanley 1994). The lagoonal lithologies and associated fauna defined therein are, in turn, succeeded by marine sandy facies younger than about 3000 years old, especially at and near core tops at the two archaeological sites. These observations are noted in cores ST3, ST7 and ST9 at Heracleion, and cores ST13 and ST17 at East Canopus.

4.2 Dating changes from lagoonal to marine settings

Palaeontological analysis of some cores highlights the change from lagoonal to marine settings that accompanied the eventual submergence of the sites in the late Holocene. Many sections of borings show a gradual up-core change from lagoonal to marine fauna, interspersed at times with locally stronger marine or freshwater influences. Among representative cores are ST9 (at Heracleion) and ST13 (at East Canopus) that record the long-term sequence of lagoonal-to-marine deposits that formed the substrate beneath the two cities (Figure 4.9). Faunal composition of ST9 records a progressive trend from lagoonal to marine settings, starting from the core base (dated 4100±40 years BP), where a freshwater influence is noted, to intermediate lagoonal–marine conditions, and to marine at the core top (540±40 years BP). Core ST13 records the same palaeoenvironmental trends: from lagoonal (from 3770±40 to about 2120±40 years BP), to intermediate lagoonal–marine (up to 1980±40 years BP), and finally to marine settings.

The up-core faunal changes are generally parallel in the two dated cores. However, it appears that the major late Holocene biofacies transitions, from lagoonal to lagoonal–marine and then to fully marine, occurred several centuries earlier at Heracleion than at East Canopus. This is based on correlation between cores and interpolation of radiocarbon-dated sections. At Heracleion, the major transitions occurred at ~2300 uncalibrated (= uncal.) years BP (lagoonal to lagoonal–marine) and ~2100 years BP (lagoonal–marine to marine). At East Canopus, similar up-core transitions are dated to ~2150 and ~1800 years BP. More radiocarbon-dated samples and identified archaeological levels would probably provide even more precise dating of the marked environmental and associated faunal changes from Greek and Roman to Byzantine times.

4.3 Problems of correlation and 'old Holocene' dates in some core sections

The lagoonal to marine sequence recorded by this study would in fact be expected in Holocene delta margin sections that, in time, were lowered from near sea level elevation to shallow marine depths in what is now Aboukir Bay. Several problems should be considered, however. The first pertains to the difficulty in making core-to-core correlations, even between borings collected near each other; such correlations cannot be achieved using only core depth, lithology and/or fauna (Figures 4.2, 4.3). Radiocarbon dates collected for these cores should provide a critical database to make stratigraphic correlations. In many cases, however, the radiocarbon ages (uncalibrated dates) vary rather incoherently from core to core. For example, they range from as young as 540±40 to 6310±60 years BP within the upper 100 cm of core tops at Heracleion, and from 1980±40 to 5990±60 years BP within the upper 50 cm at East Canopus. Moreover, at both Heracleion and East Canopus, the substrate core sections formed of mud are surprisingly old, ranging from 7050±50 years BP at the base of core ST1 to 6570±40 yrs BP at the base of core ST14. It is recalled that both of these old dates were obtained for samples recovered at relatively shallow depths of <550 cm below the sediment–water interface. Moreover, the entire core sections in these two borings, as well as in cores ST3 and ST12, are of early to mid-Holocene age (for the most-part >5000 years BP). Only the thin surficial sandy cover at the two archaeological localities is of late Holocene to modern age (Figure 4.9).

Biostratigraphic correlation between core strata, even between closely spaced borings within one area, and interpretation of 'old Holocene date' anomalies that characterise many core sections,

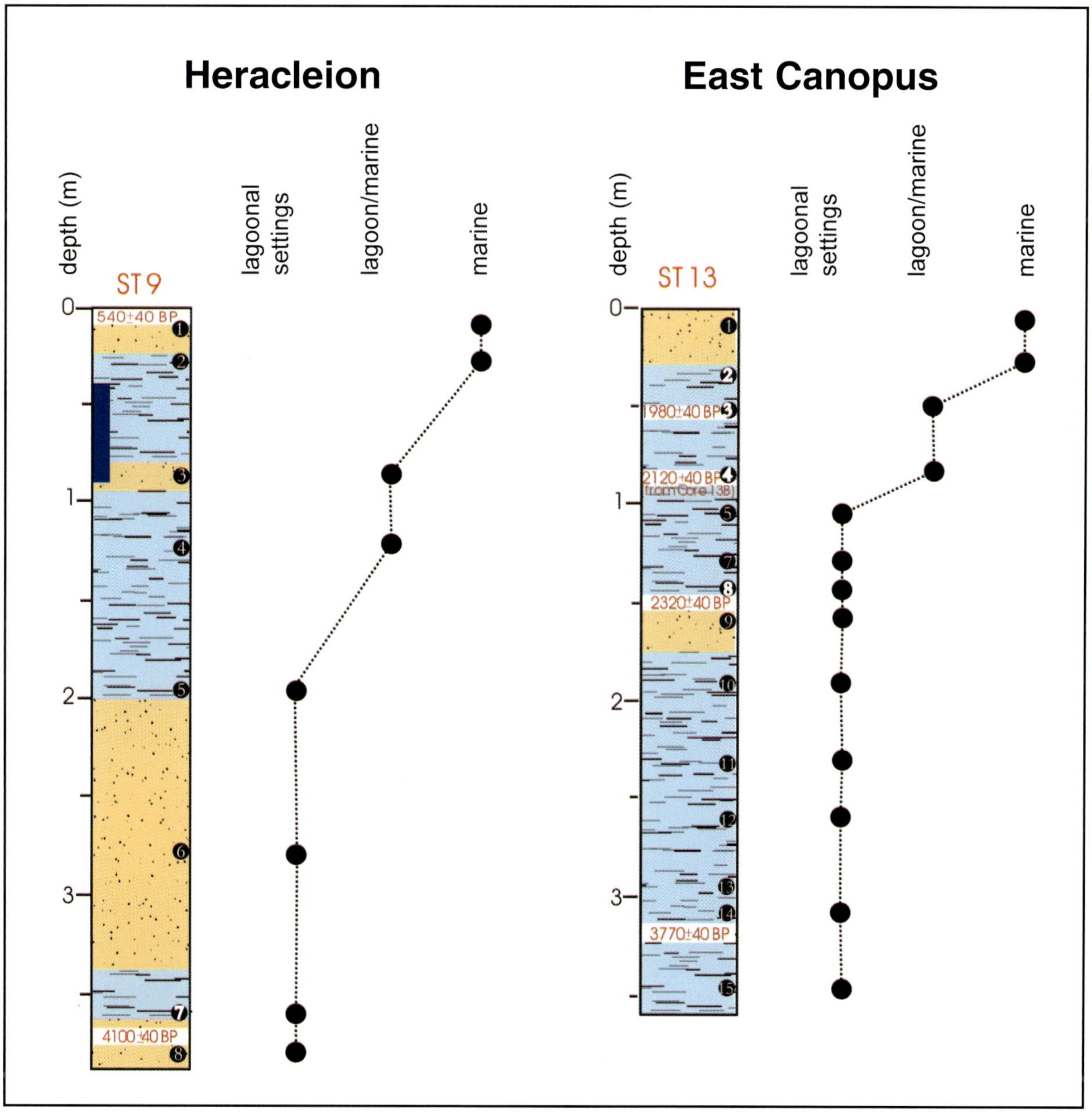

Figure 4.9 Two type-cores, one from Heracleion (ST9) and the other from East Canopus (ST 13), show fauna recording environmental changes in late Holocene time. While up-core sequences are generally similar in both, the recorded lagoonal to marine change appears to have occurred earlier at Heracleion than at East Canopus (see discussion in text). The blue band in the upper part of ST9 denotes sediment section comprising disturbed bedding.

should take into account available geological data. Pertinent among these are observed syn- and post-depositional disruptions of unconsolidated Holocene strata that now lie beneath the two cities. Evidence of stratal offset is recorded in geophysical profiles obtained in western Aboukir Bay (Stanley *et al.*, Chapter 3 in this volume). High-resolution seismic profiles, side-scan sonar images and nuclear resonance magnetometry records all show that sediment substrates beneath both cities have been offset, tilted, uplifted and/or lowered (Stanley *et al.*, 2001, 2004, and Chapter 3 in this volume). For example, diapirs, crown cracks, growth faults and normal faults have deformed the bedding that originally was laid down horizontally at and seaward of Canopic distributary mouths. X-radiographs of cores also show evidence of syn- and post-depositional deformation, primarily in the form of fluidised sediment associated with stratal offset. For example, three of the cores we examined at Heracleion (ST3, ST7 and ST9) include sections 30–50 cm thick of such disturbed beds (Figure 4.2). All such failed and displaced bedding phenomena are typical of water-saturated delta-front and prodelta lithofacies that accumulate rapidly at the sedimentologically unstable coastal margin of many large deltas (Morgan 1970; Wright and Coleman 1974; Prior and Coleman 1982; Coleman 1988; Maestro *et al.* 2002).

4.4 Difficulty in detecting evidence of sediment failure in some lagoonal successions

The geophysical surveys of Aboukir Bay indicate that the sudden large-scale events (including flooding) occurring in this region induced syn- and post-depositional failure of strata beneath the cities (Stanley *et al.*, Chapter 3 in this volume). However, at some core sites the environmental evolution through time as recorded by faunal assemblages is, in some cases, characterised by progressive, generally gradual, upward changes. Thus, in some cores, it is difficult to reliably identify corresponding abrupt up-core faunal changes induced primarily by sediment failure and lowering of strata. This lack of obvious faunal change is probably due to the fact that fauna in brackish water settings (such as lagoon and marsh) typically tolerate frequent and often important changes of salinity, temperature and turbidity. This means that changes of brackish water fauna due to chaotic and sudden sediment failure events may not necessarily be obvious. If there were no geophysical records to indicate stratal disturbance and the probability of rapid foundering of sediment in the Aboukir Bay study area, then the faunal changes due to such events could well be overlooked, i.e. masked and interpreted as the result of 'just one more' form of environmental instability in what are normally highly variable ecological settings.

4.5 Poorly preserved and stained fauna as significant markers

Additional insight on the role of events that affected substrate environments of the ancient cities is provided by core sections that contain no fauna or, in the case of molluscs, include only poorly preserved shell fragments. Moreover, with regard to microfossils, the number of samples that comprise variable proportions of oxidised (brown orange, yellowish brown) foraminiferal and ostracod specimens is noteworthy. We believe the various normal to poor states of faunal preservation provide additional insight on environmental changes, including those induced by grain size attributes and others due to stratification disturbance during and following deposition.

In the case of molluscs, for example, the grain size of samples with no fauna was compared with that of samples comprising well-preserved fauna. Samples devoid of molluscs (n in Figures 4.2 and 4.3) or 'residue' are typically characterised by small mean grain size and high proportions of silt and clay (Table 4.1). It is also noted that samples without molluscs were collected from four of five core sections older than 5700 years BP. Microfossils are absent in deposits older than 6000 years BP in cores ST12 and ST14 (East Canopus), and in sands at the base of core ST7 (Heracleion). In the case of samples with only mollusc fragments, those at Heracleion are sand-rich (fine to medium grain size) and older than about 3000 years; such samples at East Canopus are of silt-size sediment with almost no sand, and date from about 3000 to over 6000 years BP. The absence of molluscs or presence of only poorly preserved shells may be related to *in situ* conditions of burial, in settings where reducing conditions once prevailed. This is not unusual in organic and plant-rich mud of marsh and inner lagoon deposits (Ho and Coleman 1969; Loizeau and Stanley 1993). Reduced or anoxic burial condition may also explain the absence of microfossils in some core samples. Deterioration of shell material may also have been induced by vertical shifts in groundwater level and the chemical reactions related thereto. This latter case may perhaps record phases when sediment sections were subject to marked changes in water content, including strata affected by vertical shifts relative to groundwater and sea level, and those modified by effects of syn- and post-depositional fluidisation.

Also of note are core samples where at least two

of the three fauna contain one or more of the following attributes: mollusc shells that are fragmented or absent, and/or microfossils that are absent or iron-stained. This is recorded in 21 (53%) of 40 samples from the cores at Heracleion, and in 11 (33%) of 33 samples at East Canopus. There is no evident correlation of such samples with core depth and textural attributes: only 29% of the above anomalous preservations occur in sediment of unusual textural attributes at Heracleion, and 36% at East Canopus (Table 4.1). Cores ST1, ST3 and ST9 at Heracleion and core ST13 at East Canopus comprise the largest proportion of samples with stained microfossils. It is postulated that core samples with anomalous fauna are more closely related to factors other than texture, such as age and specific geographical location of the sediment section.

Some oxidised tests may be allochthonous, for example those displaced after initial deposition, such as at times of flood in the distributaries or by wave and coastal current erosion and transport along the delta coast (cf. studies in Morgan 1970). However, some staining of microfossils formed *in situ* should not be ruled out. According to Samir (2000), high organic matter present in some mud-rich sediment can lower redox potential, thus controlling the concentration of ferrous iron and limonitic colour of foraminifers and ostracods' valves. It is envisioned that some staining developed preferentially in delta margin environments where oxidation prevailed (Ho and Coleman 1969; Morgan 1970; Broussard 1975), as in well-exposed natural levees that once bordered the Canopic channels (Chen *et al.* 1992; Warne and Stanley 1993) and in high marsh sectors positioned for some time above water level.

If staining is due to *in situ* oxidation, preferentially in settings positioned above water level, it may then follow that stained fauna serve as key markers of post-depositional change. Further examination of stained fauna in the Aboukir Bay study area are needed to test the hypothesis that such fauna are closely associated with sediment strata that were periodically shifted and tilted either upward or downward relative to sea level.

5 Conclusions

The results of geological and geophysical investigations in Aboukir Bay, presented elsewhere in this volume, indicate that lowering of Heracleion and East Canopus beneath the waves was caused by gradual processes coupled with a series of more sudden and powerful events. These have involved (1) a relative rise in sea level due to the late Holocene eustatic rise and concurrent lowering of land at the delta margin by sediment compaction and isostatic loading, and (2) a sequence of more rapid-to-sudden failures of strata at and near Canopic river channel mouths (Stanley *et al.*, Chapter 3 in this volume). It has been suggested that these latter more rapid, and at times catastrophic, events were largely due to positioning of the cities in close proximity to river mouth areas subject to high water and sediment discharge, especially at times of annual Nile floods (Popper 1951; Said 1993). Cities were constructed close to the coastline directly on vulnerable delta-front deposits that were water saturated and prone to conditions of soil instability and periodic failure. It is probable that such sediment failure was most commonly triggered by fluvial events (especially during and shortly after times of unusually high flooding in late summer and fall), as indicated by Stanley and others (2001, 2004, and in this volume), and perhaps also by some episodic earthquake-related tremors and associated tsunamis (cf. Pirazzoli 1986; Stiros 2001).

The record of biofacies change (Figure 4.9) indicates that subsidence occurred earlier at Heracleion (between ~2300 and 2100 uncal. radiocarbon years BP) than at East Canopus (~2150 and 1800 uncal. radiocarbon years BP). It is expected that further refinement of these dates will be made possible by discovery of archaeological markers gathered at the underwater sites.

Up-core biofacies changes suggest that substrates beneath the two settlements were indeed lowered from primarily lagoonal to fully marine settings in a fairly brief period of time. Some disturbed mud-rich deposits are noted at Heracleion, such as in core ST9 (upper part; see Figure 4.9), ST3 (upper part) and ST7 (middle part). Fauna in Holocene sequences encountered in these cores record a series of environmental changes that could be interpreted as a response to natural wetland variations, and not to subsidence. Palaeontology records lagoonal and marine assemblages, with the marine fauna (especially molluscs) recording reduced tolerance to environmental changes. Largely marine intervals in borings indicate relative environmental stability, since the marine fauna are not able to sustain large-scale environmental variability. Lagoonal fauna of brackish water origin, on the other hand, are distributed over longer Holocene time intervals below the marine ones, and capable of tolerating more extensive variations of environmental factors. If chaotic and sudden events took place during the time of deposition of lagoonal sediments, the effects of disruption of strata may not be obvious. In fact, this effect may be partially masked by the generally tolerant nature of fauna living in the brackish water settings. However, other criteria may indicate vertical shifts of land relative to sea level. The locally present and poorly preserved molluscan fauna and iron-stained microfossils are tentatively viewed here as markers of lagoonal sediment that was subject to displacement relative to sea level.

The present biofacies study highlights some difficulties in correlating Holocene core sections using only biofacies, core depth and lithology. This is made readily apparent by the presence of highly variable radiocarbon dates (lower to late Holocene ages) at core tops and difficult stratigraphic correlation between cores owing to marked offset of Holocene strata. Better insight on the final history of the cities and aspects of their lowering during the past two millennia will be gained by means of further integration of faunal analyses with results from geological and geophysical investigations. A next step will be to determine statistically the distribution of molluscs, ostracods and foraminifers in various deltaic-related settings at the two modern Nile river mouths (Rosetta and Damietta). Such data should be compared to the results of our faunal analyses in Holocene core sections in western Aboukir Bay that once lay at the mouth of the Canopic branch. This will serve to identify and evaluate more reliably the influence of subsidence and sea level rise on biofacies that occurred in Aboukir Bay over 2000 years ago.

Chapter 5

Vibracores in Aboukir Bay: description and analyses

Thomas F. Jorstad and Jean-Daniel Stanley

'For the nature of the land of Egypt is this: as you approach it and are still within one day's run from land, and you drop a sounding line, you will bring up mud, though you are in eleven fathoms' depth. This shows that the deposit of earth reaches even as far as this.' Herodotus, *The Histories*, 2.5

A coring programme was conducted in April and May 2001 to collect surface sediment samples as an integral part of the interdisciplinary exploration and archaeological effort of the IEASM on ancient submerged cities in western Aboukir Bay (Figure 5.1). The major goal of the programme was to augment the previously collected geological information, including bathymetric, gravimetric and side-scan exploration data (obtained largely in 1996 and 1997 by previous missions of the IEASM) and high-resolution sub-bottom seismic survey profiles (collected in April and May 2000). The recovered cores are a major source of information to supplement data from the few previously described sediment borings in Aboukir Bay (Frihy 1992a,b). They also complement core logs available for the terrestrial margin of the bay, including those depicted in earlier works by Attia (1954), Stanley (2003, 2005a,b), Stanley *et al.* (1996, 2004), and unpublished engineering reports.

When used collectively, core data contribute critical information pertaining to the Holocene geology and depositional history at Heracleion and East Canopus which can refine our understanding of the regional cultural history. The intent of the core analysis was to investigate sedimentological, petrological, artefactual, faunal, floral and chronostratigraphic aspects of sedimentary sections forming the substrate beneath the two ancient cities. Examination of the substrate allows enhanced resolution of the migrating Holocene strandline and identification of lithofacies to identify the changing floodplain, lagoonal and other deltaic regimes that prevailed in this region through time.

With this information, lateral stratigraphic correlations and palaeogeographical reconstructions can be derived to interpret more accurately the original (pre-subsidence) setting of East Canopus and Heracleion (Stanley 2003; Stanley and Warne, Chapter 2 in this volume). Cores are also needed by geologists, archaeologists and other specialists to help answer questions relating cultural history to gradual land subsidence, sudden substrate failure, sea level rise and climate change. In sum, detailed analysis of subsurface sediment layers provides a better understanding of the natural history of the setting upon which the two Greek settlements were built, as well as the conditions and events that led to their demise.

1 Core recovery and accession

Coring operations were conducted in the western part of Aboukir Bay to collect continuous stratigraphic sections at both Heracleion (cores 3–5 and 8–11) and East Canopus (cores 13–17). Seventeen core sites were concentrated primarily in and adjacent to archaeological areas, and also in sectors proximal to these two submerged localities (cores 1, 2, 6, 7, 12). The specific site localities were chosen on the basis of previously collected bathymetric and high-resolution seismic profiles (Stanley et al., Figures 3.10, 3.11, 3.12, 3.15 in this volume). The drill sites, from east to west (Figure 5.1), were recovered in water depths ranging from 2.1 to 6.6 m (see detailed core logs in Appendix).

The vibracore drilling method was selected because it typically provides continuous sediment sequences in excess of 5 m with a minimal disturbance of material. The system used was a Rossfelder P-3 apparatus operated from a tender with a hydraulic crane (see Stanley *et al.*, Figure 3.16 in this volume). The core positions were identified using the Global Positioning System (GPS). During drilling operations, the boat remained stationary at each site by using several anchors secured to the sea-floor. A diver helped place the anchors and also ensured that the core barrel remained vertical during drilling.

The core drilling operations used aluminium tubing with barrels 6.1 m (20 ft) long and 10 cm (4 inches) in diameter. After drilling, some tubes containing the sediment were cut aboard ship to remove the upper empty tubing without sediment; the sediment-filled core barrels were then sealed and stored horizontally. The combined total recovery length of all 17 cores is 68.41 m. Individual cores ranged in length from ~1 m to 5.5 m, and these sections included, for the most part, continuous and complete core recovery.

The cores were cut to predetermined lengths (typically 1.2 m) using a table saw equipped with a 3 mm wide diamond blade. Next, the core sections were cut lengthwise. To minimise any contamination or alteration of core material in this procedure, the lengthwise cut of the core tube penetrated only the aluminium and did not cut into the sediment. The lengthwise separation of sediment sections was made using a metal wire that was drawn to split each section into equal halves. Each core half was wrapped in a plastic sleeve and sealed with electrical tape at each end. These sealed materials were placed in cold storage until described and sampled. All core segments are stored face-up in the cold storage facility and kept at 6° C.

2 Core log description

2.1 General methodology

The cornerstone for interpretation and analysis of the substrate setting is lithological description and compilation of petrological features that characterise the sediment core sections. A detailed lithological log was made, based on the visual and physical inspection of each split core section, with the aid of X-radiographs. As in previous stratigraphic work in the Nile delta (Stanley *et al.* 1996), all core material was carefully analysed centimetre-by-centimetre for sediment lithology (texture, composition), physical and biological structures, disturbance of sediment sections (fluidisation, microfaulting) and colour (Munsell Color 1975). In addition to the petrology of sediment layers, attention was paid to any archaeological, rock fragment, nodule, botanical and faunal materials present. Descriptions were performed in a clean-room setting with daylight and/or photoflood light conditions near the refrigerated storage area to minimise thawing or chemical and biological changes of the sediment. Selection of core material for radiocarbon dating was a major aspect of the

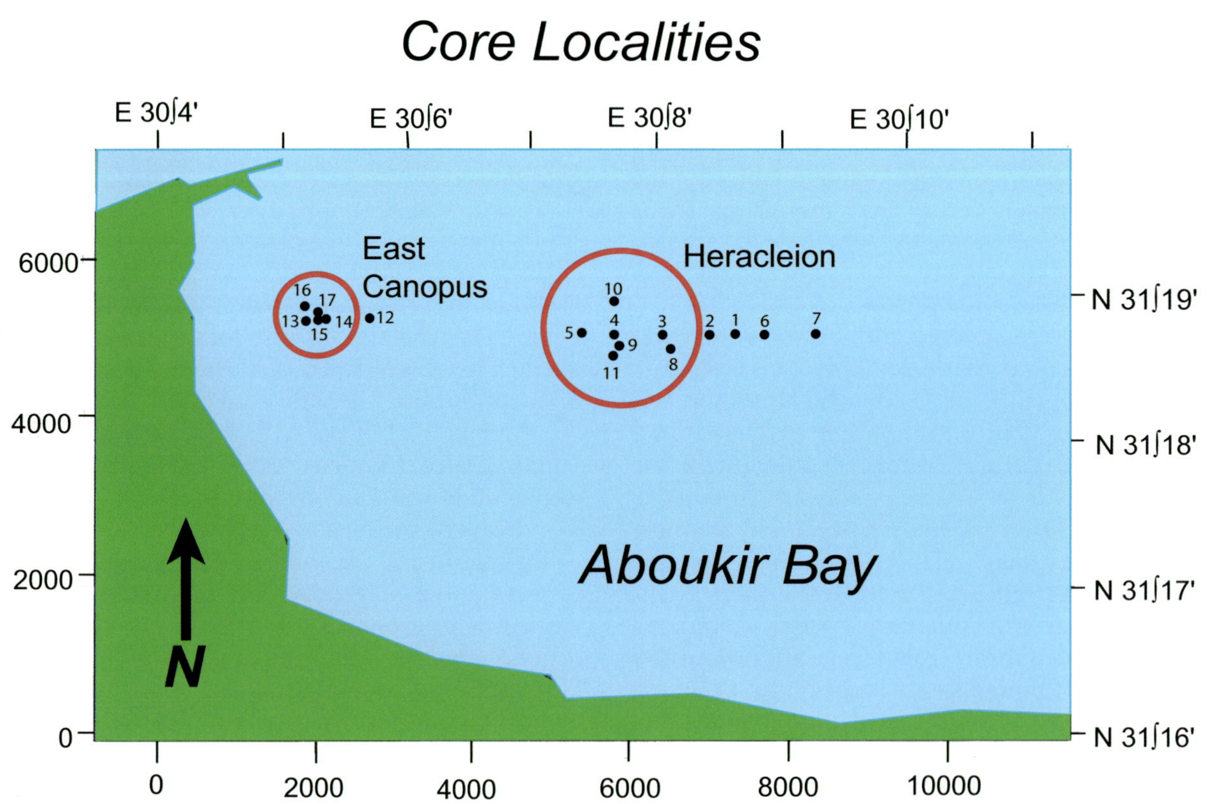

Figure 5.1 Map showing locations of the 17 vibracores collected at and near the ancient cities of Heracleion and East Canopus in Aboukir Bay.

study, and samples from borings at 15 sites in the bay were collected for this analysis. Information contained on the graphic logs also includes the geographical location in latitude and longitude (based on GPS readings) and water depth. Graphic logs, including legend and descriptions of the 17 vibracores (numbered 1 to 17), are presented in the Appendix. At several sites, two or more cores were collected (3, 5, 8, 13, 16) and their logs are also depicted (see Appendix). The reader is directed to discussions and interpretations of vibracore petrology presented by Stanley *et al.* (2001, 2004; and Chapter 3 in this volume).

2.2 Texture and composition

An analytical tool used to assist in the lithological descriptions is the Coulter Counter LS200 Particle Analyzer. Selected samples (79 to date) from nearly all Aboukir cores were analysed to calculate the proportions of particle sizes ranging from clay (<4 μm) and silt (4–63 μm) to sand (63–2000 μm). This provided a continuous suite of textural data from the base to the top of the cores. Differential and cumulative size curves were obtained for each sample. Also collected for each sample were percentages of sand, silt and clay, mean and modal size, and standard deviation (listed in Bernasconi *et al.*, their Table 4.1 in this volume). This analysis statistically identifies the major textural trends through time in each of the cores, and also serves to determine similarities and differences between cores. These observations, in turn, can assist in the interpretation of the depositional setting and archaeological history of the substrate in the study area.

Information on the composition of sediment is obtained by means of binocular microscopic examination of bulk samples, and also by thin sections viewed with a petrographic microscope. Identification of diverse rock (lithic) samples serves to interpret local bedrock and more distal source terrains, and sheds light on the transport process of bedload materials in former river channels and other depositional conditions.

2.3 X-ray documentation

X-ray photography is an excellent method to extract additional information in sediment that is typically invisible to the naked eye, such as subtle sedimentary and biogenic structures. X-radiographs were taken of each core section shortly after the aluminium tubing was sectioned and split. Extensive experimentation was conducted prior to documentation of the Aboukir core material. Numerous X-ray systems and film types, both film-based (Eastman Kodak Company 2000) and digital formats, were tested in order to provide the finest detail and resolution in the recovered sediment. In addition, the use of Computed Tomography (CT) to scan structural detail in selected sediment sections was also investigated. Smithsonian Institution collaboration in the testing phase of the X-ray work involved the NMNH Departments of Paleobiology, Anthropology, and Vertebrate Zoology, as well as the Smithsonian's Center for Materials Research and Education.

Based on this experimentation, a totally new application using a health imaging diagnostic film, EktaScan EM-1, was created for use on inorganic material. All X-ray images were completed using a Porta Ray model #400 with a 37 inch focal length. Core samples were exposed at 80 kilovolts (KV) for 2.5 seconds. Both the archive and the sample halves of the core slices were photographed together using 35 x 43 cm (14 x 17 inches) sheet film. The film was automatically processed using a M35A X-Omat Processor. The X-ray system and processing unit were located in the NMNH's Department of Anthropology. Examples of sedimentary structures visible using X-ray photography are depicted by Stanley (2003, 2005b) and Stanley *et al.* (Figure 3.18 in this volume). For each core segment, there is a digital X-ray record stored in CD-ROM format along with a series of archival volumes with hard-copy prints.

2.4 Radiocarbon sampling and analysis

The purpose of developing a radiocarbon-dating based chronology was to better distinguish various key subsurface Holocene sedimentary units for stratigraphic mapping and correlation. With such a chronostratigraphic framework in place, archaeologists can then correlate artefactual horizons, create timelines for cultural history and improve excavation strategy. During the process of describing core lithology, all potentially datable material (some as small as ~2 mm in size) was collected from the split core sections, placed in vials and recorded. Sixty-two samples, including charcoal, wood, peat and a variety of plant material, as well as shell, were separated as potential material for age determination.

Depending on sample size, weight and material used, either Accelerator Mass Spectrometry (AMS) or conventional radiometric technique was used with standard pretreatment and sample preparation. The two methods of dating samples provide the same age result, but use different types of detectors (liquid scintillation counter for conventional runs on moderate to large samples, and accelerator mass spectrometer for very small samples). All the

Table 5.1 Radiocarbon dating results and associated information for vibracore and surficial samples in Aboukir Bay.

Site	*SI ID*	*Field ID*	*Lab No.*	*Sample Code*	*Depth from TOC (cm)*	*Sample Type*
AB	0022UB3883		145896	GODDIO#1		Conventional
AB	0022UB3883		145897	GODDIO#2	TW2	AMS
AB	0022UB3883		145898	GODDIO#3	trench sample	AMS
AB	0022UB3883		145899	GODDIO#4	zone115-15	AMS
AB	0022UB3883		145900	GODDIO#6	TW1	AMS
AB	0022UB3883		145901	GODDIO#7	TW4	Conventional
AB	0022UB3883		145902	GODDIO#10	trench	AMS
AB	0022UB3883		145903	GODDIO#11	"core, lower"	AMS
AB	0022UB3883		145904	GODDIO#12	TW4	Conventional
AB	0022UB3883		145905	GODDIO#13	T2000	AMS
AB	0022UB3883		145906	STANLEY SITE T		AMS
AB	0022UB3883		145907	STANLEY T ZONE		AMS
AB	1A I		169951	AQ191-93CM	91-93	AMS
AB	1A III		169964	AQ1250-270	250-270	Conventional
AB	1A VI	PF1	159415	AQ1A505-515	505-515	AMS
AB	1A VI	PF1	159416	AQ1A545-549	545-549	AMS
AB	3B II	PF3A	159417	AQ3B92	92	AMS
AB	3B III	PF3A	159418	AQ3B175	175	AMS
AB	3B III	PF3A	159419	AQ3B223-227	223-227	AMS
AB	4A I		169952	AQ452-65	52-65	AMS
AB	4A II	PF4	159420	AQ4A212	212	AMS
AB	4A III	PF4		AQ4A299-304	299-304	AMS
AB	5A II	PF6B-2	159422	AQ5A155-159	155-159	AMS
AB	5A IV	PF6B-2	159423	AQ5A282-287	282-287	AMS
AB	5B II	PF6A	159822	AQ6AII138cm	138	AMS
AB	5B II	PF6A	159823	AQ6AII175cm	175	AMS
AB	7A I		169953	AQ7115-122	115-122	AMS
AB	7A II		169954	AQ7160-170	160-170	AMS
AB	8B I	PF8B	159821	AQ8B63-66	63-66	AMS
AB	9A I		169955	AQ921-24	21-24	AMS
AB	9A IV	PF9-2	159424	AQ9A382-390	382-390	AMS
AB	10A II	PF10	159425	AQ10A136-141	136-141	AMS
AB	11A I		169971	AQ110-7	0-7	AMS
AB	11A I		169956	AQ1122-40	22-40	AMS
AB	11A II	PF11	159426	AQ11A156-158	156-158	AMS
AB	11A II	PF11	159427	AQ11A195-198	195-198	AMS
AB	11A III	PF11	159428	AQ11A292-297	292-297	AMS
AB	12A I		169965	AQ1237-57	37-57	Conventional
AB	12A II		169966	AQ12100-120	100-120	Conventional
AB	12A II	PF12-1		AQ12A112-114	112-114	AMS
AB	12A III		169967	AQ12202-222	202-222	Conventional
AB	12A IV	PF12-2	159430	AQ12A308	308	AMS
AB	13A II	PF13-2	159431	AQ13A65	65	AMS
AB	13A II		169957	AQ13152-157	152-157	AMS
AB	13A IV		169958	AQ13323-329	323-329	AMS
AB	13B I	PF13B	159825	AQ13BI36-39	36-39	Conventional
AB	13B I		168608	"AQ-13B, 68-72"	68-72	AMS
AB	13C I		168609	"AB-13C,18cm"	18	AMS
AB	13C I	PF13A	159824	AQ13AI32-36	32-36	Conventional
AB	14A II		169959	AQ14128-137	128-137	AMS
AB	14A III		169968	AQ14244-264	244-264	Conventional
AB	14A V		169960	AQ14450-455	450-455	AMS
AB	15A I		169961	AQ1550-54	50-54	AMS
AB	15A I		169972	AQ15112-115	112-115	Conventional
AB	15A III		169969	AQ15270-290	270-290	Conventional
AB	16B I	PF16B	159826	AQ16BI3-4.5	3-4.5	AMS
AB	17A I		169962	AQ1741-51	41-51	AMS
AB	17A I		169963	AB1797-102	97-102	AMS
AB	17A II	PF17-2-1		AQ17A157-158	157-158	AMS
AB	17A II		169970	AQ17170-186	170-186	Conventional
AB	17A IV	PF17-2-2	159433	AQ17A355-359	355-359	AMS

Date Sampled	Date Sent	Measured Radiocarbon Age (BP)	13C/12C Ratio ‰	Conventional Radiocarbon Age (BP)	Sample Description
	14/7/00	2270 ± 60	-26.2	2250 ± 60	wood
	14/7/00	2370 ± 40	-27.8	2330 ± 40	plant material
	14/7/00	1980 ± 40	-23.4	2010 ± 40	charred seeds
	14/7/00	3140 ± 40	-25.6	3130 ± 40	charred material
	14/7/00	2030 ± 40	-23.4	2070 ± 40	wood
	14/7/00	6500 ± 70	-29.1	6440 ± 70	peat
	14/7/00	2060 ± 40	-12	2270 ± 40	wood
	14/7/00	5530 ± 50	-19.7	5620 ± 50	wood
	14/7/00	6760 ± 60	-25.8	6750 ± 60	peat
	14/7/00	2090 ± 40	-26.6	2070 ± 40	wood
	14/7/00	2300 ± 50	-23	2330 ± 50	wood
	14/7/00	2390 ± 40	-26.8	2360 ± 40	plant material
	19/8/02	5340±40	-0.01	5750±40	shell
	19/8/02	6150±110	-19.3	6240±110	organic sediment
21/8/01	5/9/01	6640 ± 60	-26	6620 ± 60	plant material
21/8/01	5/9/01	7050 ± 50	-26.4	7030 ± 50	plant material
22/8/01	5/9/01	6250 ± 60	-21.6	6310 ± 60	plant material
23/8/01	5/9/01	5650 ± 40	-14.5	5820 ± 40	plant material
23/8/01	5/9/01	6590 ± 40	-22.1	6640 ± 40	plant material
	19/8/02	3330±60	-18.7	3430±60	organic sediment
27/8/01	5/9/01	5440 ± 50	-20.3	5520 ± 50	plant material
27/8/01	5/9/01	no date			plant material
28/8/01	5/9/01	5980 ± 60	-14.9	6150 ± 60	plant material
28/8/01	5/9/01	5550 ± 40	-19.4	5640 ± 40	plant material
12/9/01	20/9/01	5880 ± 50	-21.1	5940 ± 50	plant material
12/9/01	20/9/01	5530 ± 50	-14.9	5700 ± 50	plant material
	19/8/02	2630±40	-16.1	2780±40	plant material
	19/8/02	3260±40	-2.4	3710±40	shell
13/9/01	20/9/01	6330 ± 50	-21.4	6390 ± 50	plant material
	19/8/02	120±40	0.9	540±40	shell
29/8/01	5/9/01	3980 ± 40	-17.6	4100 ± 40	plant material
30/8/01	5/9/01	NA	NA	6300 ± 60	plant material
	19/8/02	1140±40	1.4	1570±40	shell
	19/8/02	3180±40	-19.5	3270±40	organic sediment
30/8/01	5/9/01	5100 ± 50	-29.1	5030 ± 50	plant material
30/8/01	5/9/01	6240 ± 40	-24.6	6250 ± 40	plant material
30/8/01	5/9/01	5990 ± 40	-19.3	6080 ± 40	plant material
	19/8/02	5890±60	-18.7	5990±60	orgamic sediment
	19/8/02	5840±60	-18.9	5940±60	organic sediment
31/8/01	5/9/01	no date			plant material
	19/8/02	6090±60	-19.3	6190±60	organic sediment
31/8/01	5/9/01	5590 ± 40	-18.2	5700 ± 40	plant material
31/8/01	5/9/01	1980 ± 40	-24.8	1980 ± 40	plant material
	19/8/02	2320±40	-24.9	2320±-40	plant material
	19/8/02	3350±40	0.8	3770±40	shell
13/9/01	20/9/01	2240 ± 70	-26.4	2220 ± 70	plant material
	31/7/02	2140 ± 40	-26.5	2120 ± 40	plant material
	31/7/02	2230 +- 40	-26.3	2240 ± 40	plant material
14/9/01	20/9/01	2070 ± 70	-26.5	2050 ± 70	plant material
	19/8/02	5780±40	-19.2	5880±40	organic sediment
	19/8/02	5900±60	-19.1	6000±60	organic sediment
	19/8/02	6200±40	-2.6	6570±40	shell
	19/8/02	2420±40	3	2880±40	shell
	19/8/02	2660±50	1.7	3100±60	shell
	19/8/02	5640±70	-18.8	5740±70	organic sediment
14/9/01	20/9/01	6920 ± 60	-27.2	6880 ± 60	plant material
	19/8/02	2440±40	0.5	2860±40	shell
	19/8/02	2930±40	0.1	3340±40	shell
2/9/01	5/9/01	no date			plant material
	19/8/02	5160±60	-20.4	5240±70	organic sediment
3/9/01	5/9/01	5610 ± 40	-19.5	5700 ± 40	plant material

radiocarbon analyses were performed by Beta Analytic Inc. in Miami, Florida.

The database compiled for all radiocarbon-dated samples is listed in Table 5.1, and is also shown in stratigraphic context on the graphic core logs. The database records sample ID information, sample depth, type of dating method employed, sample material used, and the measured and conventional radiocarbon dates. Of the 62 total samples collected for dating, 61 were submitted for processing and, of those, 3 samples had insufficient material for dating. Plant matter (n = 51), including charcoal, carbonised wood, leaf, seed and peat, and shell (n = 10) were the available materials selected for dating. Twelve of these samples (Lab numbers 145896 to 145907, in Table 5.1; see also Appendix 2 for calibrated radiocarbon dates), recovered from the surficial horizon of the substrate at East Canopus, are included in this inventory. The results of dating analysis are discussed by Stanley *et al.* (Chapter 3 in this volume) and Bernasconi *et al.* (Chapter 4 in this volume).

2.5 Sampling for other analyses

Faunal analyses

Numerous samples were collected for faunal analyses from the core material. Some of these contained one or more large molluscan shells for specific study. Most (n = 79), however, were bulk samples from which representative microfaunal assemblages (foraminifera, ostracods) and molluscs (gastropods, pelecypods, scaphopods) were extracted from the sediment matrix for statistical biofacies study (Bernasconi *et al.*, Chapter 4 in this volume). Faunal analysis serves to interpret the changing palaeogeography of the substrate, from early Holocene to the present, and provides information on changing coastline conditions and positions before, during and after subsidence of the archaeological sites.

Botanical analyses

In addition to collecting floral material from the vibracores for radiocarbon dating (see graphic logs), many botanical specimens, including peat material, seeds, wood, charcoal and grasses, were sampled and are under study for interpreting depositional environments. The initial analyses for identification of these materials include photographic documentation (traditional and digital formats) and Scanning Electron Microscopy (SEM). In addition to comparative analysis using floral research collections, studies have been initiated to identify the material using genetic and DNA methods with the assistance of specialists at the NMNH and in other organisations.

Geochemical analyses

Bulk sampling of selected cores was made to investigate aspects of environmental geochemistry (core 13), including palaeopollution (from mid-Holocene to present). For this, isotopic measurements are being made of heavy metals, including lead. In addition, strontium isotopic analyses for salinity and other measurements are also being made on core 13. Moreover, a toxicological characterisation of the clayey sediment at the two archaeological sites (cores 7 and 17) is currently in progress.

3 Perspectives

As demonstrated in other chapters of this volume, investigations of the 17 recovered vibracores have provided valuable information that assists in the interpretation of the morphological and high-resolution sub-bottom profiles collected in the now submerged western sector of Aboukir Bay. Particularly useful is the evidence indicating that Holocene sections at Heracleion and East Canopus have been subject to syn- and post-depositional processes and failure of sediment substrates (Stanley *et al.* 2003, 2005a,b; Stanley *et al.* 2004, and Chapters 3 and 4 in this volume). The presence of early to mid-Holocene sequences at the top of cores and the marked offset nature of Holocene sections from core site to core site highlight the importance of large-scale substrate failure beneath both ancient cities. Core analysis has also served to distinguish some natural from anthropogenic causes of site damage and submergence.

A number of questions need to be more fully answered and, of these, perhaps the most pressing pertain to the nature and timing of lowering of the archaeological sites beneath the waves. The botanical record will help refine substrate characteristics, such as elevation relative to sea level, and help determine former drainage conditions. This type of information may provide additional insight as to whether submergence occurred progressively or suddenly, or by a combination of both processes. Moreover, calculations of the rate of bay submergence could be made by more precise correlation between the vibracores recovered in the bay proper and cores collected in the terrestrial sectors along the bay margin (Frihy 1992a,b; Stanley *et al.* 1996, 2004). The still not fully exploited geochemical data from cores, including analyses of selected elements and heavy metals, may provide insight on the economic nature of human activity (metal and glass works, ceramics) when the cities were still occupied.

The information from the cores, in conjunction with other hard physical data, will continue to serve as a crucial database to help make coherent geoarchaeological interpretations of the now submerged ancient sites.

Appendix 1

Logs of Vibracores in Aboukir Bay

Core Number: ABUQ 1A
Location: N 31° 18.76' E 30° 8.63'
Water Depth: 6.2m

Abu Qir/Alexandria 2001

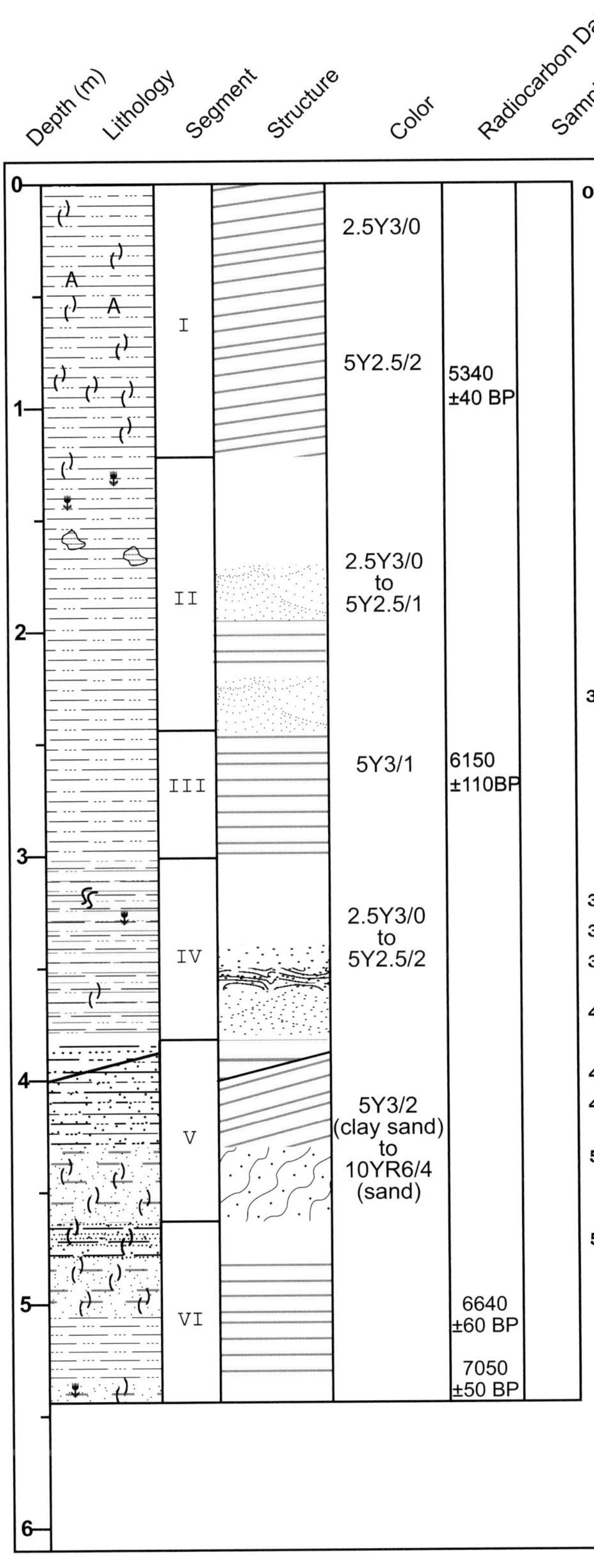

0-301cm: Silty clay, soft, plastic, siliceous.

0-122cm: Dispersed shell fragments and whole shell visible (see X-ray also).

0-60cm: Somewhat laminated in places, laminae at ~15∞ angle.

42cm: Sample: artifact, small grooved lithoclast.

50cm: Sample: artifact, irregular shaped siltstone.

60-80cm: Distinct lamination apparent on X-ray, laminae at ~15∞ angle.

80-122cm: Somewhat laminated in places, laminae at ~15∞ angle.

90-95cm: Shells.

122-144cm: Rather homogeneous, somewhat micaceous; fine root filaments(?).

162cm: Small siltstone piece, 1x2cm.

166cm: Irregular siltstone piece, ~4cm diameter.

170-244cm: Laminae and cross-bedding; units horizontal, ~2cm thick, some wavy laminae.

244-301cm: Micaceous; faintly to moderately laminated; a few parting surfaces stained 2.5Y4/2.

301-382cm: Clay and silty clay, soft and plastic.

321cm: Lithified worm tube.

326cm: Some very decomposed plant fragments observed under reflecting light microscope.

340-380cm: Cross-lamination apparent in X-ray.

350-360cm: Appears disturbed in X-ray.

362cm: Some small shell fragments.

382-392cm: Silty clay and silt; well laminated.

392cm: Truncation appears in X-ray, angle ~25%.

392-430cm: Fine sand and silty clay; well laminated, lamination at 10-15∞ and occurs in 2-3cm units, commonly with sharp base and top.

430-463cm: Shelly clayey sand; bioturbated.

450-463cm: Many whole shells observed.

463-480cm: Sand, silt, clay, shelly.

480-505cm: Clayey sand, shelly; sand laminated, laminae to 3cm thick; some articulated whole shell.

505-540cm: Silty clay, rather massive looking on X-ray; faintly laminated.

505-515cm: Sample: C14, AMS, small plant fragments collected under reflecting light microscope.

540-549cm: Dense clayey sand. Some wood and shell fragments; shells observed near base.

545-549cm: Sample: C14, AMS, small wood fragments; in better condition than sample from 505-515cm.

Core Number: ABUQ 2A
Location: N 31° 18.76' E 30° 8.43'
Water Depth: 2.6m

Abu Qir/Alexandria 2001

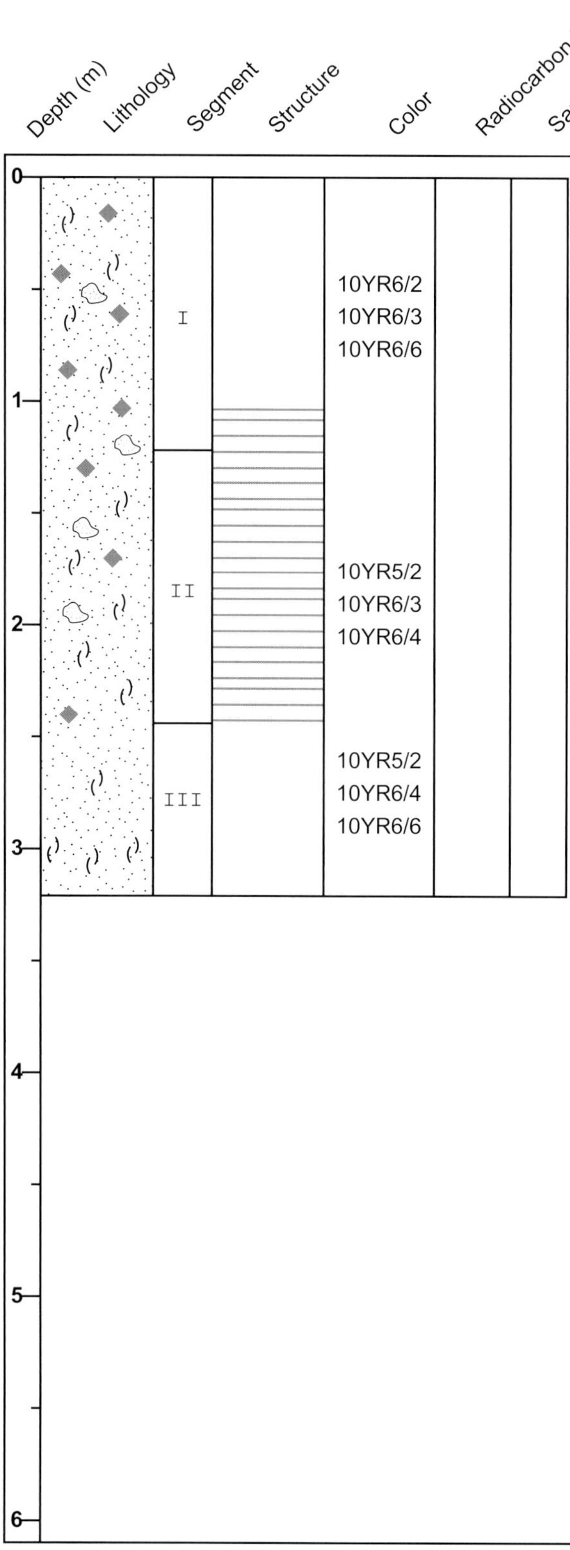

0-244cm: Fine to medium grained sand, loose, siliceous. Abundant shell fragments and shells, mostly bivalve mollusks, many quite large; many shells appear fresh, although rounded shell and/or concretions common, concretions about 1-4cm diameter and rounded; concretion less common downward, but shell fragments abundant. Sandstone lithoclasts are quite common throughout, but do not show up on X-ray.

100-244cm: Subtle bedding in places seen on X-ray.

159-199cm: Color mottling; mottles 7.5YR7/4.

244-321cm: Fine sand, siliceous; wetter and denser, fewer shells and shell fragments.

265-280cm: Mottled, 7.5YR7/4.

310-321: Well-preserved bivalve shells.

Core Number: ABUQ 3A
Location: N 31° 18.76' E 30° 8.05'
Water Depth: 5.2m

Abu Qir/Alexandria 2001

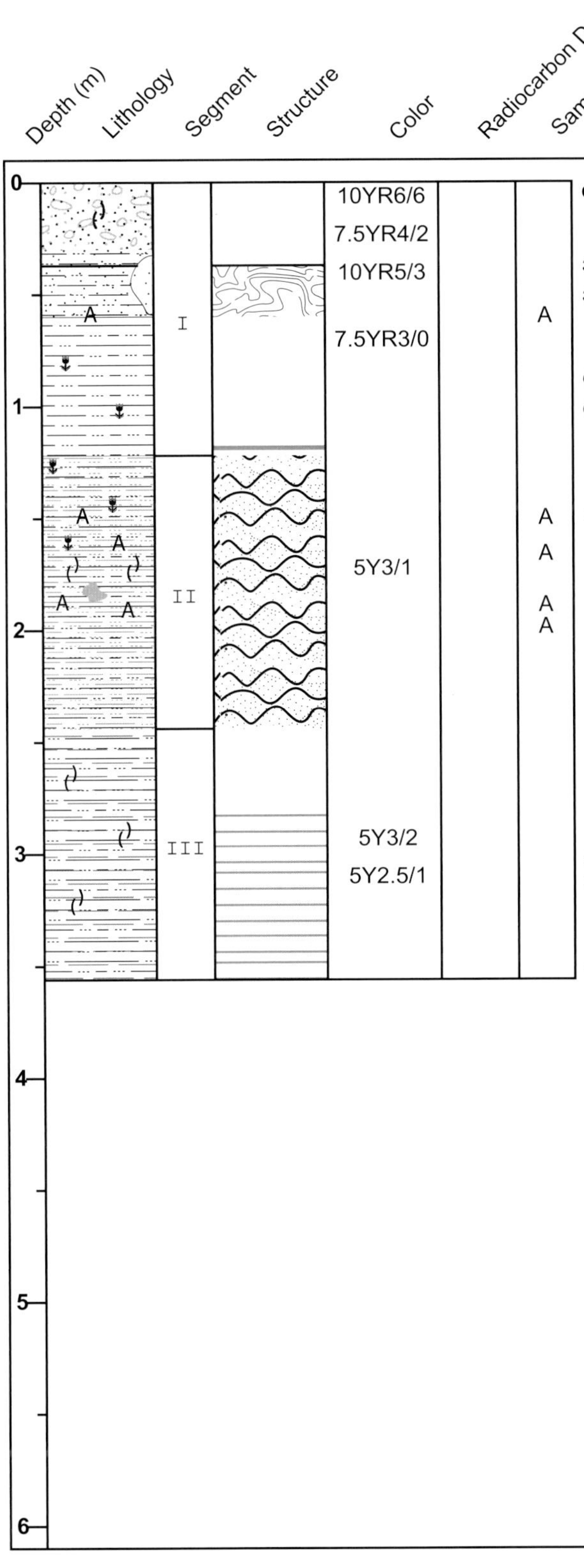

0-30cm: Sand, fine to medium grained, siliceous; minor calcareous (shell fragment) grains, some coarse grained size shell fragments; quartz grains clear, rounded.

30cm: Sharp contact, possibly disturbed.

30-60cm: Sand and silty clay; sand along side may be artifact of coring, possibly a continuation of overlying sand or a disturbed separate unit.

37-60cm: Disturbed.

60cm: Sample: artifact, ceramic, fan-shaped, 5x6cm, black, basalt(?).

60-122cm: Silty clay, siliceous, micaceous, soft, plastic; some scattered plant stems, largely decomposed, small (~5mm length). X-ray indicates distinctive rectangular object and large lithic object in archive half.

115-122: Laminated.

122-244cm: Silty clay and clayey silt, siliceous, micaceous; well laminated, laminae in units 2-3cm thick, wavy, with numerous fine laminae, hummocky.

122-180cm: Scattered plant fragments common.

150cm: Sample: artifact, small lithified consolidated piece.

166cm: Sample: artifact, lithoclast.

170cm: Lithified consolidated piece.

175cm: Shell fragments.

185cm: Lithified consolidated piece.

189cm: Sample: artifact, small lithofact, probably naturefact.

194cm: Sample: artifact, small lithofact.

220-244: Siltier, less dense.

244-356cm: Silty clay and clay, siliceous, micaceous; a few small shell fragments.

280-356: Well laminated, small 1-5mm thick laminae; wavy, but not as hummocky as above.

320-356cm: Some silt layers (10YR 5/4), ~1cm thick.

Core Number: ABUQ 3B
Location: N 31° 18.76' E 30° 8.05'
Water Depth: 5.2m

Abu Qir/Alexandria 2001

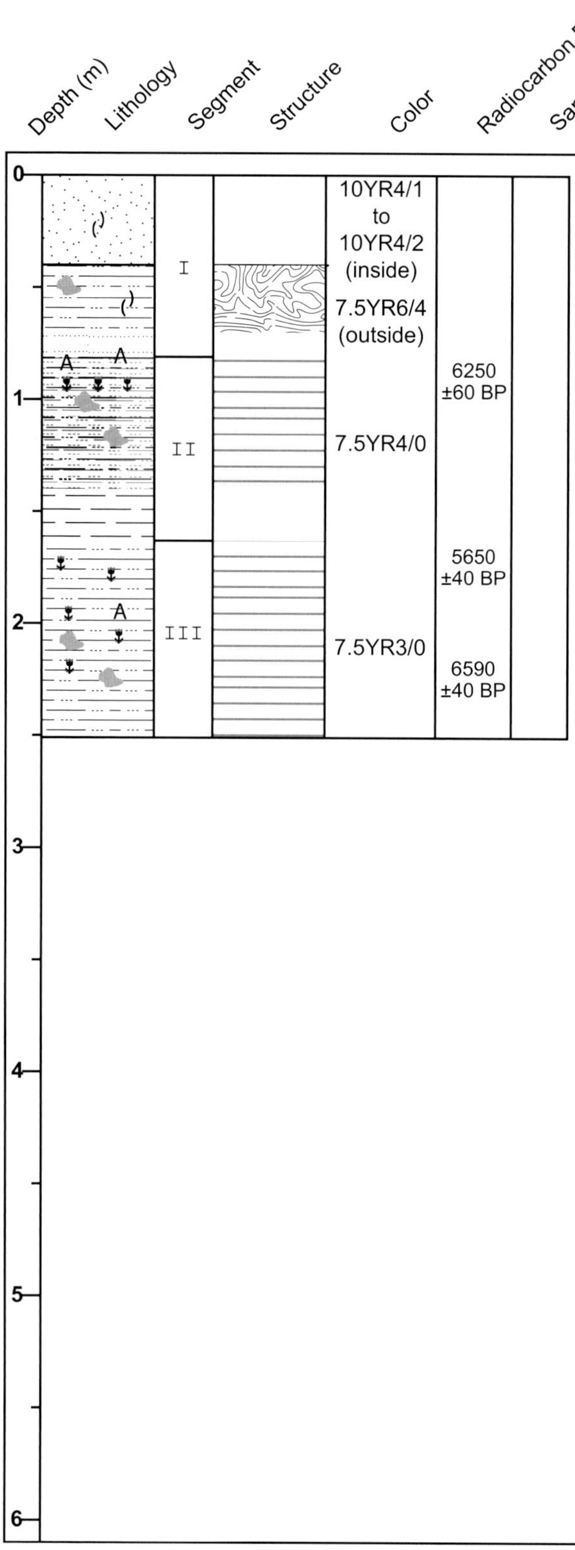

0-40cm: Sand, fine to medium grained; some 1cm shell fragments, otherwise siliceous sand. Change in sand color may be cause by exposure to air.

40cm: Rather abrupt, uneven contact.

40-70cm: Silty clay, siliceous, micaceous; scattered fossils, some well preserved; somewhat disrupted.

62cm: Sample: rock, pebble, rounded, 4cm diameter; epiphytic, calcareous, algae coat a large portion; silty around pebble.

70-81cm: Somewhat silty.

81-140cm: Silty clay and clayey silt, soft, plastic, micaceous; appears homogenous on surface and when probed, but X-rays indicate well-developed laminae.

82cm: Sample: artifact, large lithoclast.

84cm: Sample: artifact, large lithoclast, grooved.

92cm: Organic layer, wood, reed stems.

92cm: Sample: C14, AMS.

97cm: Large siltstone lithoclast, ~7cm long, 3-4cm wide, irregular, bow-tie shape with straight 1-2mm striation in middle, contains plant fragments.

100-120cm: Laminae well developed, ~2cm units.

133cm: Large silt lithoclast, ~8cm long, up to 5cm wide with distinct 1-2cm straight striation.

140-163cm: Clayey, though somewhat silty.

163-251cm: Silty clay, micaceous. X-rays show a number of dark, rounded spots (1-2cm) that can't be identified, suspected to be plant material; some discoloration (7.5YR4/2) that seems to indicate highly decomposed plant material. X-ray shows moderate laminae, uneven, ~2-3cm, generally horizontal, with gradational contacts at top and base and little internal development.

163-220cm: Many fine rootlets on broken surface.

175cm: Sample: C14, AMS, highly decomposed plant matter.

180-220cm: Some 1-1.5cm pieces of partially lithified silt found.

202cm: Sample: artifact, irregular shape, grooved in center part; color 10YR6/3 to 10YR5/2; hardened silt.

223-227cm: Sample: C14, AMS, decomposed plant material.

240cm: Obvious color band (7.5YR 4/4).

Core Number: ABUQ 4A
Location: N 31° 18.76' E 30° 7.67'
Water Depth: 4.3m

Abu Qir/Alexandria 2001

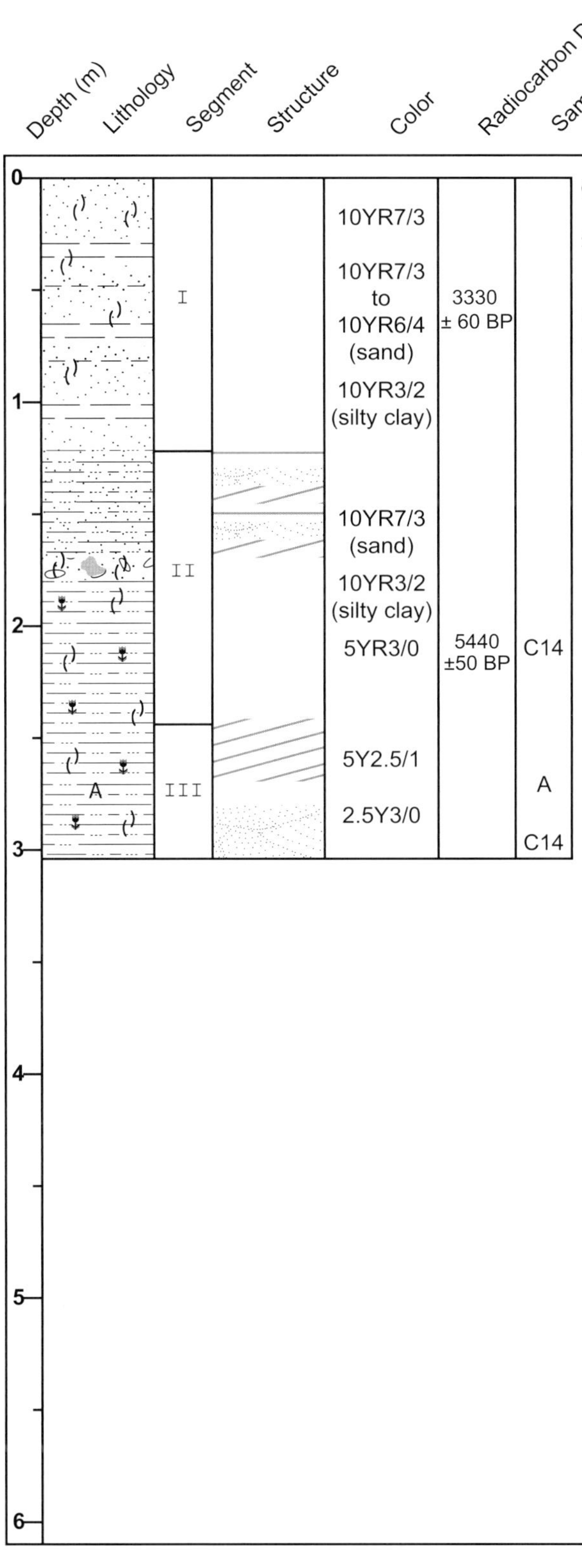

0-27cm: Sand, medium grained, siliceous; shell fragments 10-20%, black mineral 5-10%.

27-110cm: Delta front. Alternating sand, clay, silty sand; sand layers typically 3-6cm thick, rather irregular contacts; a few shells apparent in X-ray.

110-122cm: Sandier, 70-80% sand.

122-170cm: Silty clay and sand; complexly laminated, laminae ~1-2cm, discontinuous in places, wavy in places, inclined at 10∞.

170-180cm: Shelly conglomerate layer containing large beach rock clast, ~10x7x5cm.

180-304cm: Silty clay, siliceous, micaceous; breaks into plates ~1cm thick; some scattered small (~10mm) highly decomposed plant fragments and shell fragments.

212cm: Sample: C14, AMS, small plant fragments collected under microscope.

242-270cm: Somewhat laminated, laminae at 15∞.

273-279cm: Sample: artifact, large lithoclast, possibly quartzite, one end is broken, some striations, triangular, 8cm to a side, 4cm thick.

280-344cm: Some cross-lamination.

299-304cm: Sample: C14, AMS, small, highly decomposed plant fragments; not enough material for dating.

Core Number: ABUQ 5A
Location: N 31° 18.76' E 30° 7.42'
Water Depth: 5.5m

Abu Qir/Alexandria 2001

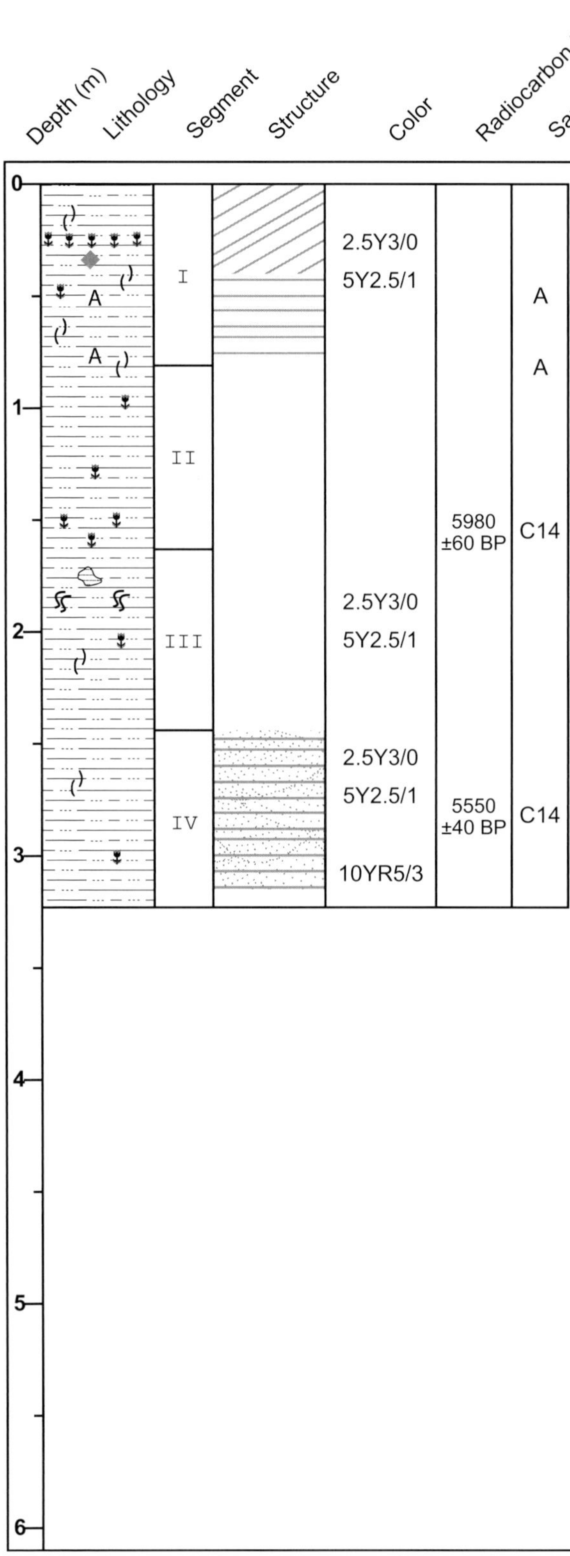

0-323cm: Silty clay, siliceous, micaceous; contains small plant fragments.

0-81cm: Very few small (less than 10mm) shell fragments.

0-40cm: Well laminated, laminae 1-2cm thick, at 30∞ angle.

29cm: Plant-bearing laminae, 0.5mm thick.

40-81cm: Slightly laminated.

40cm: Small concretion containing cordium shell.

50cm: Sample: artifact, lithoclast, large, irregularly shaped, some striations, appears to be siltstone, 5x6x3cm; color 2.5Y3/0, 5Y2.5/1.

82cm: Sample: artifact, lithoclast.

90-100cm: Plant fragments common.

150-163cm: Plant fragments common.

155-159cm: Sample: C14, AMS, moderately decomposed plant fragments.

163cm: Siltstone lithoclast cut during cutting of core barrel; appears to be typical of other siltstone lithoclasts; color 2.5Y3/0, 5Y2.5/1.

163-244cm: Appears massive in X-ray, tends to break into plates.

163-200cm: Worm tubes common, some lithified, ~0.5cm in diameter, 2-3cm long.

213cm: A few shell fragments.

244-323cm: Faintly laminated, broken surface shows some cross-lamination.

282-287: Shell fragments.

282-287cm: Sample: C14, AMS, moderately decomposed wood and plant fragments.

Core Number: ABUQ 5B

Location: N 31° 18.76' E 30° 7.42'

Water Depth: 5.5m

Abu Qir/Alexandria 2001

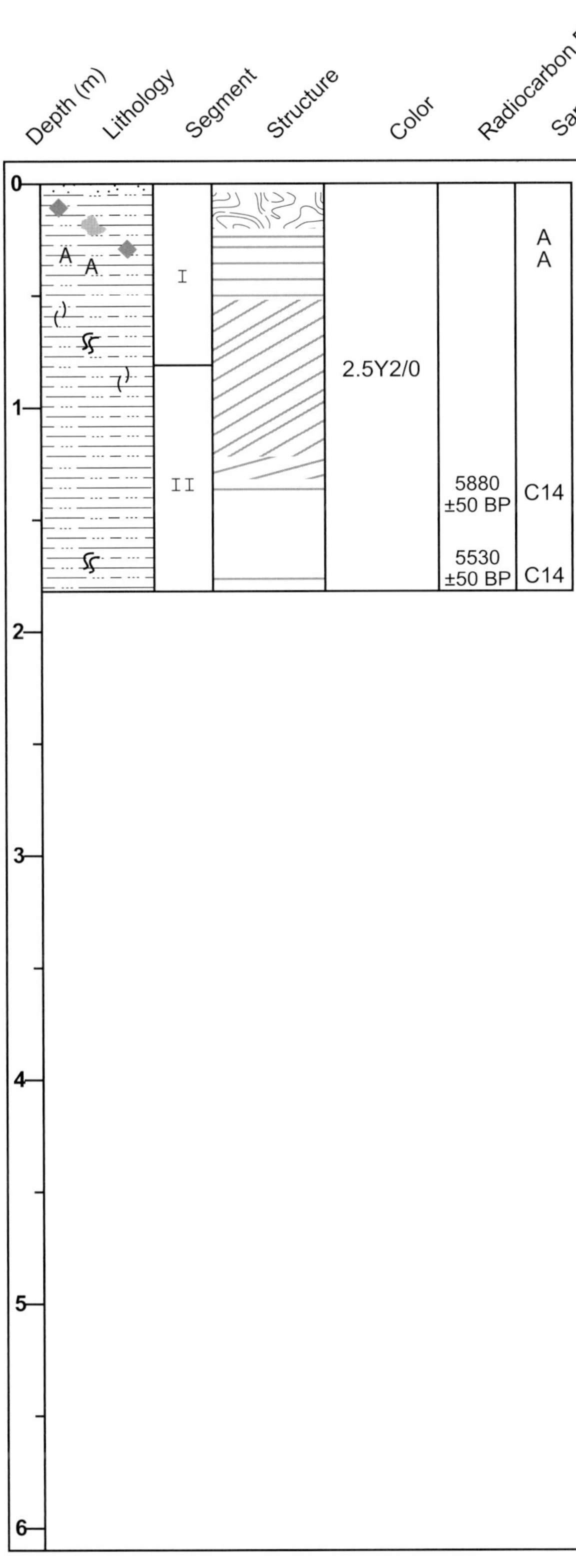

0-4cm: Fine grained silty sand; shells up to 2cm; large (up to 5cm) concretions and rock fragments with some calcareous, shelly sand nodules; mixed, irregular, contorted.

4cm: Distinct boundary.

4-182cm: Silty clay, dark, plastic, micaceous.

4-52cm: Mottling present.

4-52cm: Very finely laminated; wavy laminae 1-3mm thick; alternating light and dark blocks.

4-20cm: Mixed, irregular, contorted.

4-18cm: Large (up to 5cm) concretions and rock fragments with some calcareous, shelly sand nodules.

5cm: Possible ceramic fragment in archive half.

12cm: Flat rock fragment, 4x5cm, partly encrusted with shells; ceramic artifact.

28cm: Artifact: possible netsinker; platy rock fragment, notched in places.

30cm: Artifact: small rounded pebbles, shelly rock fragments

52-122cm: Laminated; laminae dip steeply (~30∞); alternating light and dark blocks; bounded on upper and lower margins by rock fragments (1-3cm) and articulated bivalve (1cm).

70-81cm: Thin tubules.

122-142cm: Very faint dipping strata trending to horizontal; alternating light and dark bedding. Occasional 2mm silver mica flakes.

138cm: Sample: C14, AMS, charcoal and carbonized wood material.

142-181: Massive.

175cm: Tubule or root cast, 4 cm long.

175cm: Sample: C14, AMS, abundant, non-carbonized plant material.

178-182cm: Very faint laminations.

Core Number: ABUQ 6A
Location: N 31° 18.76' E 30° 8.86'
Water Depth: 5.8m

Abu Qir/Alexandria 2001

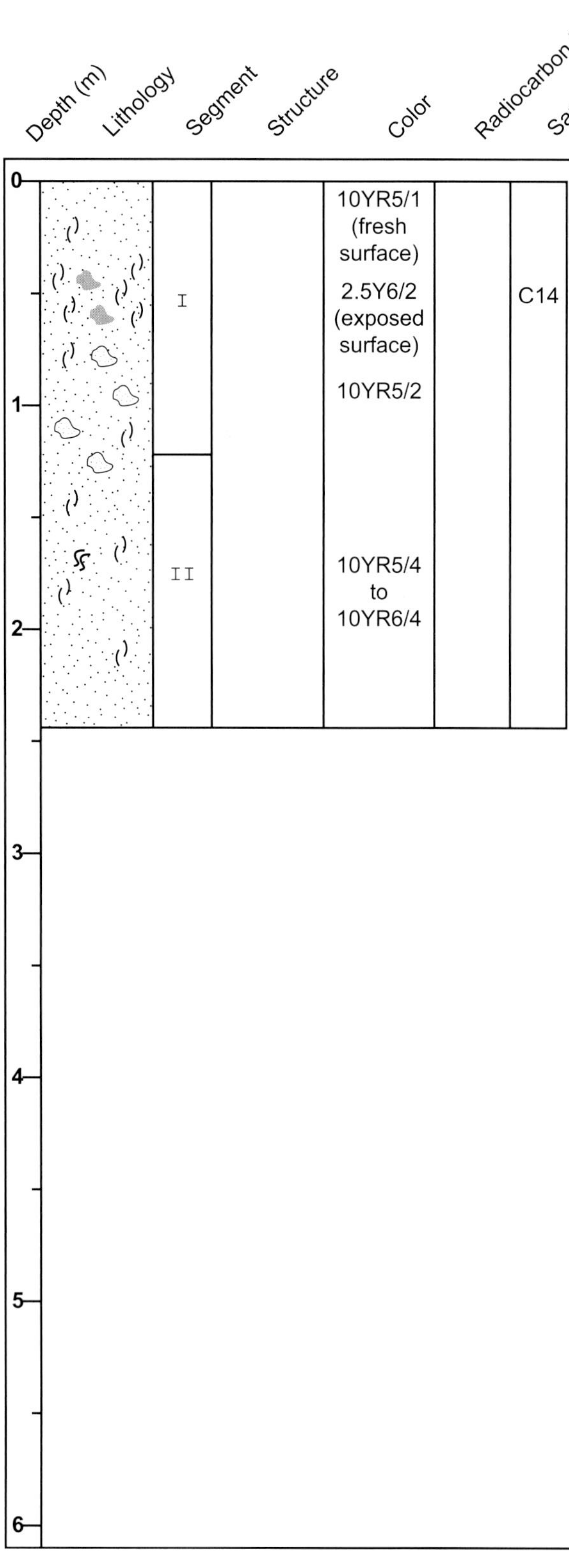

0-244cm: Sand, fine to medium grained, quartzose, 10-15% shell, 2-3% dark mineral; quartz clear, rounded; very few shell fragments larger than fine grain size.

48-63cm: Shell layer, many whole shells and well preserved bivalves; some 1-2.5cm sandstone/quartzite lithoclasts, rounded.

50cm: C14 sample taken, not sent.

80-122cm: Rounded sandstone lithoclasts, 0.5cm; seen in X-ray as small mottles.

170cm: Lithified worm tube.

Core Number: ABUQ 7A

Location: N 31° 18.76' E 30° 9.26'

Water Depth: 6.5m

Abu Qir/Alexandria 2001

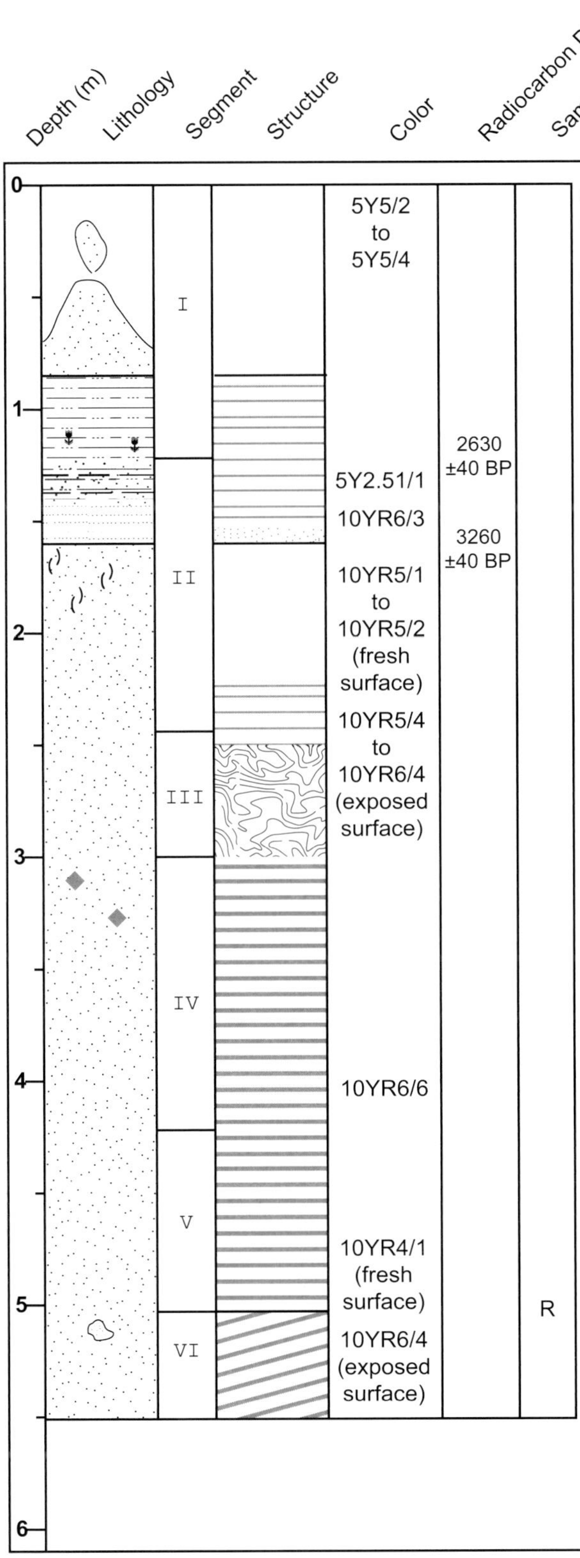

0-40cm: Empty, no recovery.

40-85cm: Sand, siliceous, fine to medium grained, only ~10% carbonate, some (2%) dark mineral; quartz clear, rounded; a few shells.

85cm: Abrupt contact.

85-122cm: Silty clay, siliceous, soft, plastic, distinctly laminated; silt laminae, ~2-3mm thick, distinct, not continuous, generally horizontal, some show cross-bedding.

115-122cm: Dispersed plant fiber.

122-140cm: Silty sandy mud, siliceous, distinctly laminated; silty laminae (10YR6/3) are very easy to discern on exposed surface.

140-143cm: Sandy mud, very soft and wet.

143-160cm: Silty laminae, well developed cross-laminae near base.

160cm: Abrupt, uneven transition.

160-551cm: Sand, medium grained, siliceous; carbonate (shell fragments) < 10%, dark mineral 2-3%, clear round quartz.

160-180cm: Some shell.

220-422cm: Horizontal laminae, fine, 2mm.

250-300cm: Laminae wavy in places, may be distortion from coring.

300-330cm: Some sandstone concretions, rounded, 0.5cm diameter, composition seems to be same as surrounding sand.

422-503cm: Laminated, 1-5cm thick.

503-551cm: Laminated, 1-5mm thick; generally horizontal, but varies from wavy to cross-bedded.

505cm: Sample: rock.

510cm: Highly crenellated sandstone concretion, 5x4x4 cm, generally same composition as surrounding sand.

Core Number: ABUQ 8A
Location: N 31° 18.66' E 30° 8.11'
Water Depth: 5.6m

Abu Qir/Alexandria 2001

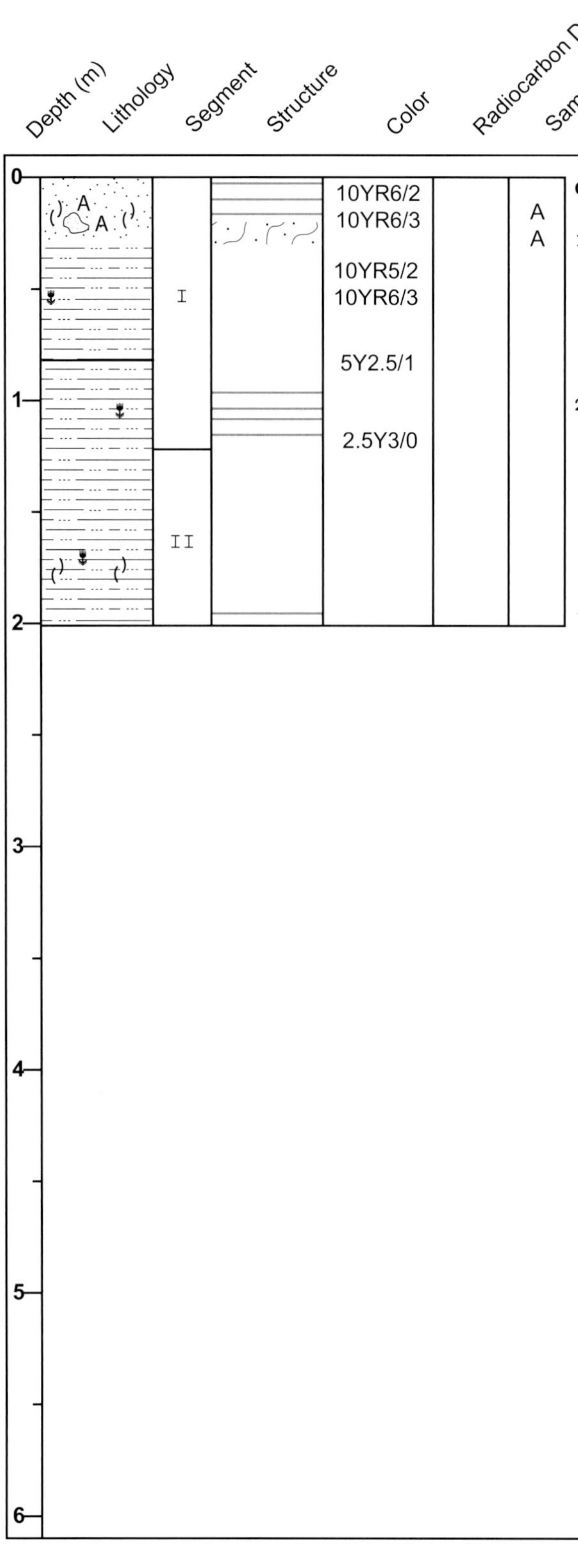

0-12cm: Sand, generally medium grained, mix of siliceous and carbonate; some dark minerals.

12-28cm: Shelly sand, medium to coarse grained.

16cm: Sample: artifact, weathered potsherd, covered with encrusting barnacles.

25-28cm: Several sandstone pebbles, partly round, 1-2cm; some are covered with bryozoa(?).

28cm: Sample: artifact, 2 small lithoclasts with indentations.

28-201cm: Silty clay, siliceous, micaceous, soft, plastic; some highly decomposed plant fragments.

30-50cm: Well developed laminae, 1-3cm.

50-60cm: Becomes bioturbated.

82cm: Hiatus.

95-122cm: Laminae moderately well developed, horizontal.

174cm: Small shell fragments.

190-200cm: Some faint 2-3cm laminae

Core Number: ABUQ 8B
Location: N 31° 18.66' E 30° 8.11'
Water Depth: XXXX

Abu Qir/Alexandria 2001

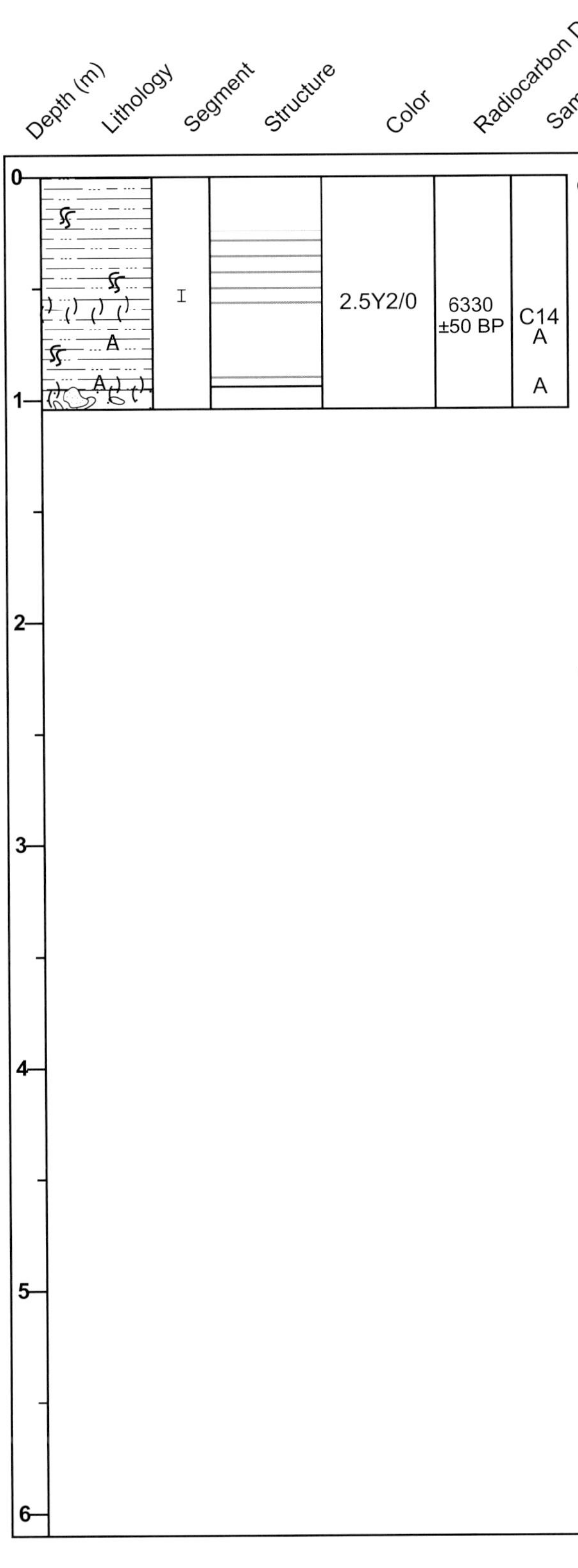

0-95cm: Silty clay, dark, micaceous, plastic; tubules filled with brown clay; very minor shell material.

24cm: Possible discontinuity.

24-63cm: Laminated.

24-42cm: Laminae 1-3cm thick, alternating light and dark.

36cm: Sample: SEM; isolated plant fragment.

42-63cm: Laminations 2-3cm thick.

63-72cm: Shell band.

63-66cm: Sample: C14, AMS; plant material, carbonized and woody.

63cm: Articulated bivalve, 1 cm; with other shell fragments.

72cm: Artifact: notched lithoclast, 2 cm; possible netsinker.

84-94cm: Very finely laminated; horizontal laminae.

88-94cm: Sand filled tubule.

94cm: Sharp discontinuity.

94cm: Artifact: netsinker; notched and grooved sandstone rock fragment, covered with shells and other organics.

95-104cm: Shell hash with large sandstone fragment and artifacts.

100-104cm: Sandstone fragment, 9x3cm, partially encrusted with shell.

Core Number: ABUQ 9A
Location: N 31° 18.68' E 30° 7.71'
Water Depth: 3.8m

Abu Qir/Alexandria 2001

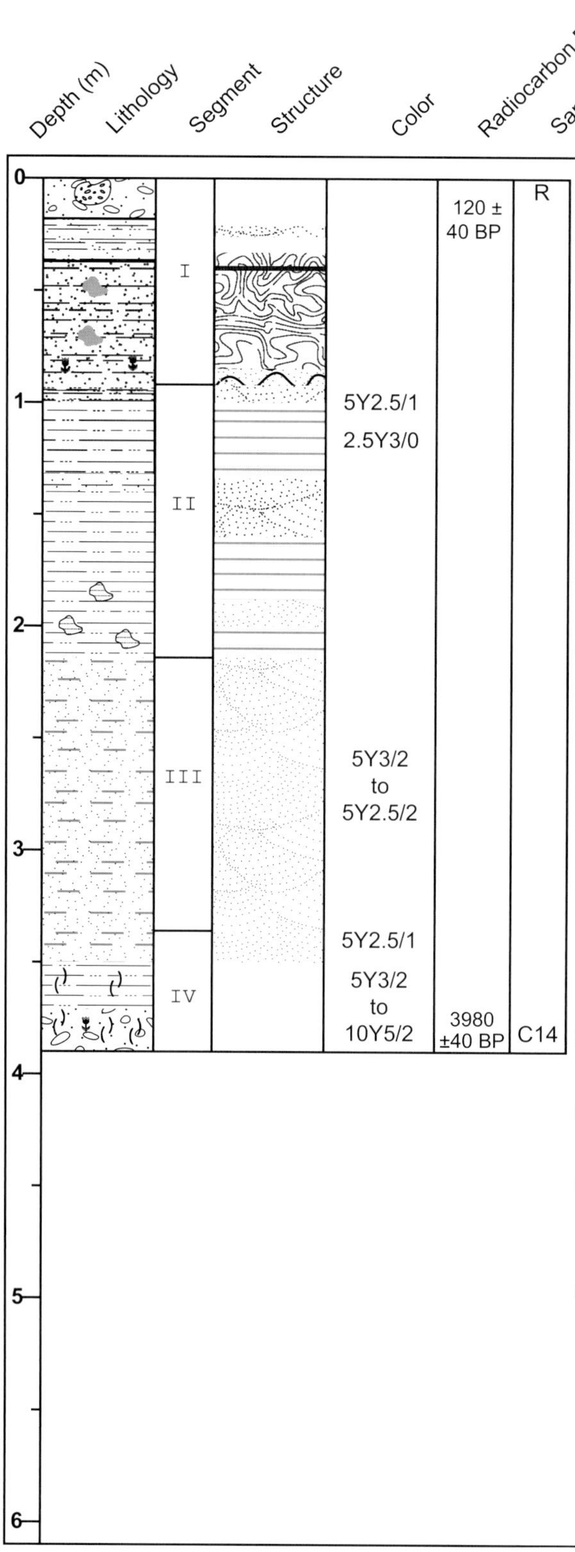

0-10cm: Large piece of kurkar ridge; approximately 10x9x9cm, 10YR8/3 in color.

0-8cm: Sample: rock.

10-18cm: Conglomerate of sand, carbonate lithoclasts, and shell fragments.

18cm: Abrupt contact.

18-37cm: Silty and fine sandy clay; siliceous, but contains shell fragments.

21-26cm: High angle cross-bedding.

33-40cm: Possible sediment intrusion structure.

37-92cm: Fine sandy clay to fine silty sand, siliceous, micaceous; X-rays show complex sedimentary structures, perhaps liquefaction.

43cm: Carbonate lithoclast.

60-92: Slightly finer grained.

80cm: Large lithoclast/artifact visible in X-ray in archive side.

85-92cm: Some highly decomposed plant fragments.

85cm: Hummocky cross-stratification apparent in X-ray.

92-100cm: Silty clay, clayey silt, and fine sandy clay, siliceous, micaceous; complexly cross-bedded.

100-132cm: Rather stiff silty clay; generally horizontal to wavy lamination.

132-134cm: Clayey silt, wet, soft.

134-188cm: Coarsening upward sequence.

134-160cm: Cross-bedding.

134-140cm: Sandy clay.

160-188cm: Well-defined horizontal laminations.

183cm: Siltstone lithoclast, 4.5x3x0.5cm.

188-214cm: Silty clay.

188-200cm: Cross-bedding.

200-214cm: Horizontal laminations.

210-214: A few siltstone concretions.

214-350cm: Fine clayey sand, siliceous, micaceous; extensively and complexly cross-bedded; hummocky cross-stratification and climbing ripples common.

350-371cm: Silty clay, shelly, siliceous, micaceous; hosted structure observed. Whole shell layer; some small scattered highly decomposed plant fragments.

371-390cm: Shell hash, sandy, some plant and wood fragments; shells commonly whole.

382-390cm: Sample: C14, AMS, moderately to highly decomposed plant and wood fragments.

384cm: Catfish spine.

Core Number: ABUQ 10A
Location: N 31° 18.98' E 30° 7.67'
Water Depth: 5.4m

Abu Qir/Alexandria 2001

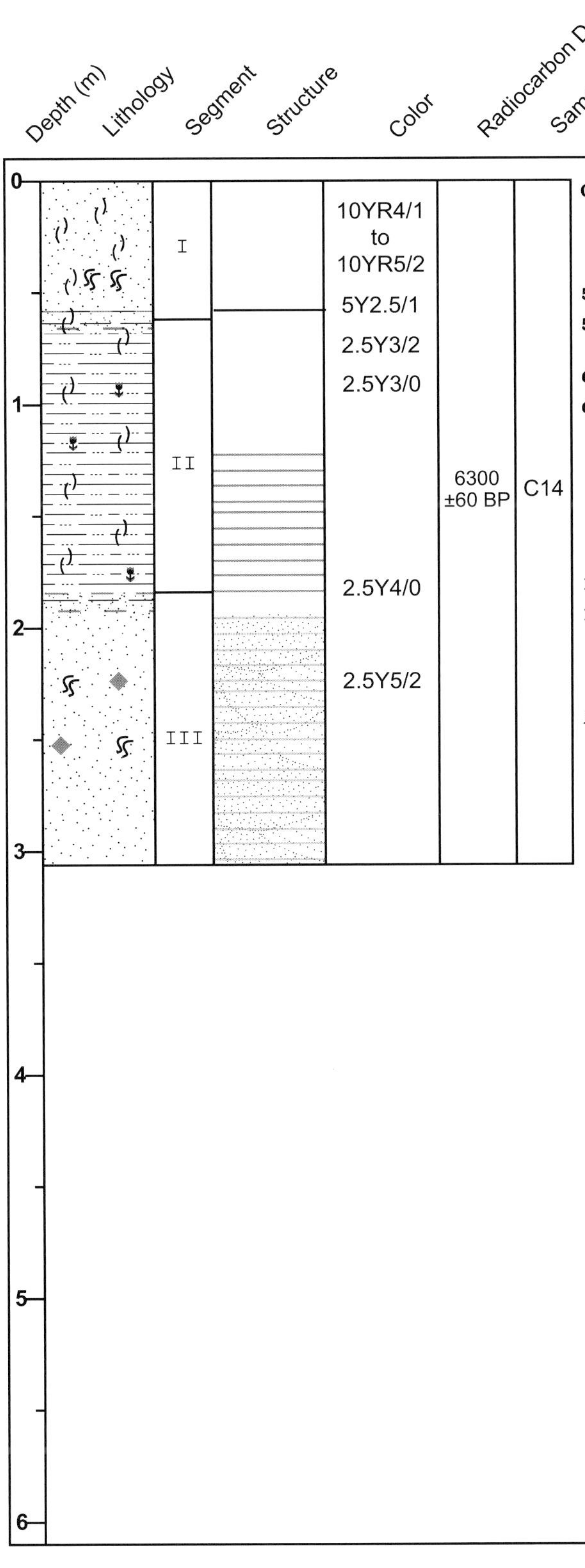

0-58cm: Sand with common shell and shell fragments, fine to medium grained; shell fragments compose ~30%, are commonly complete halves of bivalves and well preserved.

44-58cm: Worm tubes.

58cm: Abrupt contact.

58-62cm: Clayey sand, sand medium grained and contains ~30% shell fragments.

62-68cm: Clayey sand and sandy clay with shell.

68-184cm: Silty clay, siliceous, micaceous, scattered shell fragments and small, generally highly decomposed plant matter; rather stiff.

120-184cm: Indistinct laminae, ~5cm thick; on broken surface appears to be 2-3mm cross-lamination.

136-141cm: Sample: C14, AMS, moderately well preserved leaf fragments; not enough material for 13C/12C ratio measurement, so Conventional Radiocarbon Age given.

184-194cm: Muddy sand and sandy mud.

194-306cm: Fine to medium sand, siliceous, almost entirely quartz, well rounded and clear, about 2% dark minerals; broken surface shows indications of lamination and cross-bedding.

227cm: A few concretions and worm tubes.

246cm: A few concretions and worm tubes.

Core Number: ABUQ 11A
Location: N 31° 18.62' E 30° 7.67'
Water Depth: 4.0m

Abu Qir/Alexandria 2001

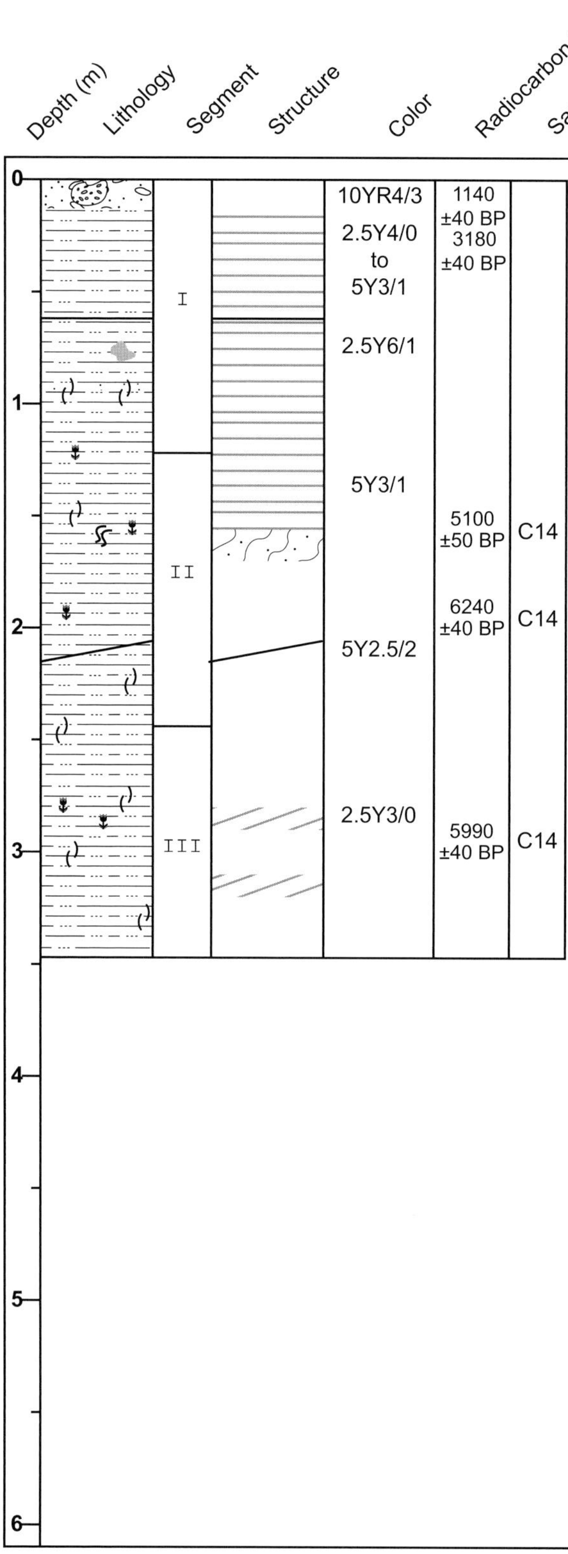

0-13cm: Conglomerate of shell fragments, mud, and shell with other carbonate fragments.

0-8cm: Weathered kurkar, 8x7x7cm.

13-347cm: Silty clay, siliceous, micaceous, soft, plastic.

13-160cm: Well laminated, laminae are 0.5-2mm thick, seen as color banding.

62cm: Discontinuity seen in X-ray.

70cm: Lithoclast, broken during coring, 8x4.5x0.5cm.

91-94cm: Sandy, shelly layer.

122-160cm: Scattered plant fragments.

145-155cm: Some horizontal burrowing, macro shells.

155-160cm: Bioturbated.

156-158cm: Sample: C14, AMS, well preserved reed stems.

160-244cm: Scattered plant fragments.

160-170cm: Bioturbated.

195-198cm: Sample: C14, AMS, rather well preserved leaves.

232cm: High angle (40∞) discontinuity in X-ray.

244-347cm: Exposed surfaces break into platy pieces; scattered fine shell.

280-290cm: Laminae at 20∞; visibly scattered plant material, generally decomposed.

292-297cm: Sample: C14, AMS, moderately well preserved plant material; small quantity.

310-320cm: Some lamination apparent, 20∞ angle.

Core Number: ABUQ 12A

Location: N 31° 18.87' E 30° 5.70'

Water Depth: 4.2m

Abu Qir/Alexandria 2001

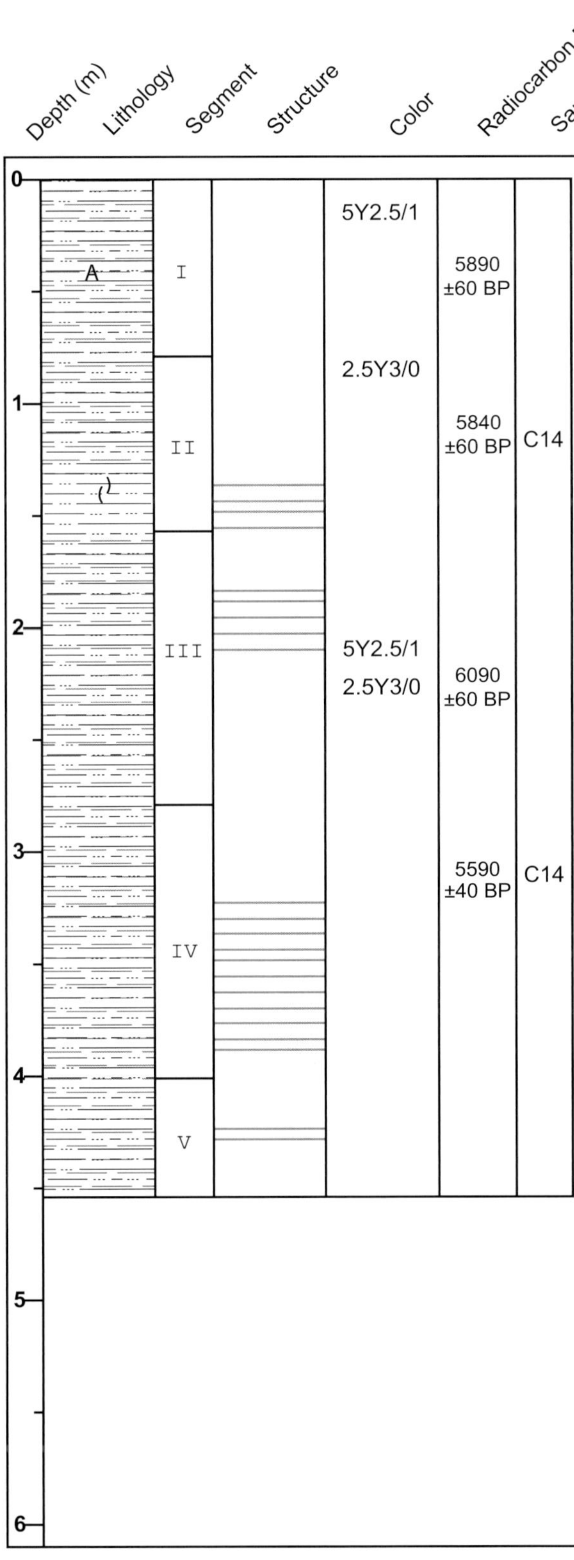

0-132cm: Silty clay and clay, siliceous, soft, plastic.

40cm: Sample: artifact, rounded lithoclast, 2.5x2.5x3cm, grooved, composed of siltstone.

112-114cm: Sample: C14, AMS; not enough material for dating.

132-157cm: Stiff silty clay; subtle horizontal lamination, 2-4cm; a few fine shell fragments.

157-454cm: Silty clay and clay, siliceous; soft and plastic.

180-215cm: Faint horizontal lamination, laminae 1cm.

308cm: Sample: C14, AMS, disseminated wood fragments; small quantity.

320-390cm: Faint horizontal laminae, laminae 1cm.

420-430: Faint lamination.

Core Number: ABUQ 13A
Location: N 31° 18.85' E 30° 5.18'
Water Depth: ~3m

Abu Qir/Alexandria 2001

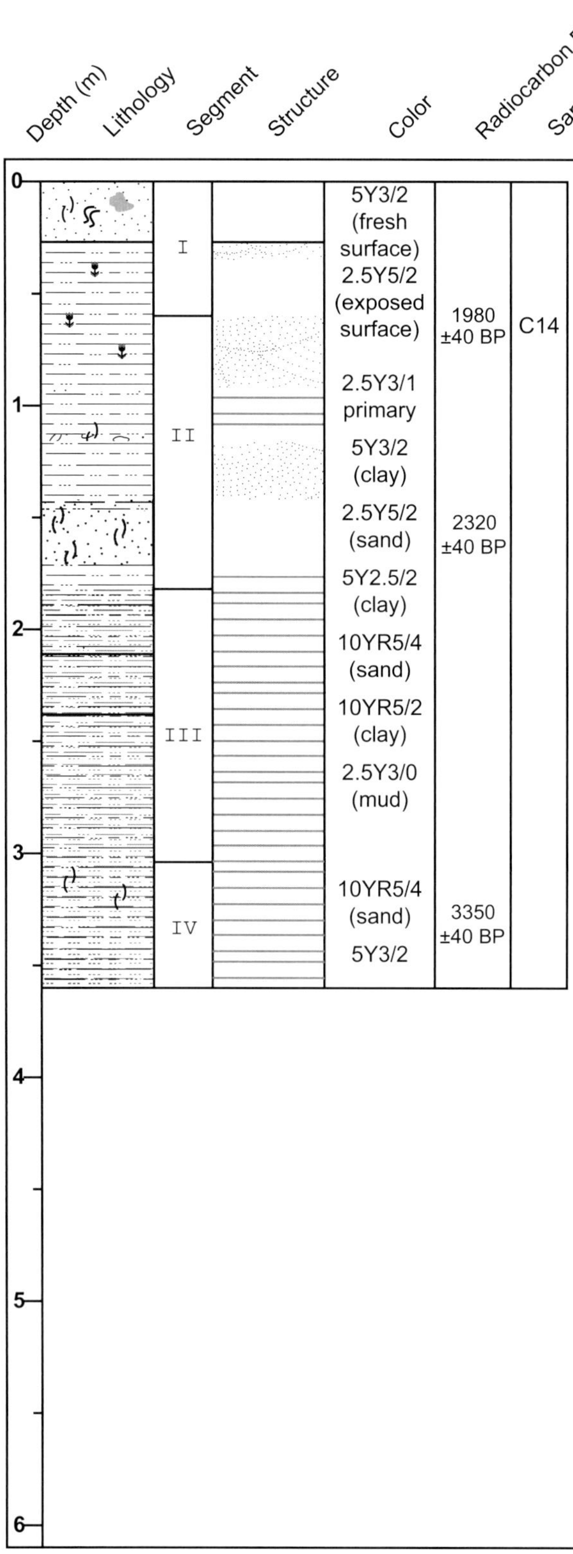

Description

0-27cm: Sand, medium grained; heterogeneous mix of quartz, shell fragments, dark minerals, mica, lithic fragments; some worm tubes.

27cm: Abrupt contact

27-92cm: Silty clay, rather stiff, micaceous.

27-35cm: Distinct cross-lamination.

37cm: Plant fragments, including possible small reed fragments.

37-45cm: Distinct color mottling, 2.5Y6/2.

55-92cm: Scattered plant material.

60-92cm: Broken surfaces show developed cross-lamination, including distinct lighter colored laminae.

65cm: Sample: C14, AMS, intact, well preserved, flattened(?) vasicular stem; candidate for SEM.

92-94cm: Sand, fine grained.

94-113cm: Silty clay, micaceous, laminated, laminae 2cm.

104cm: Distinct 3mm laminae, 2.5Y6/2.

113-116cm: Sand, mix of quartz, shell fragments, dark minerals.

116-142cm: Silty clay, well developed cross-lamination.

143-146cm: Mud and sand, very soft.

146-148cm: Silty clay.

148-172cm: Sand, medium to coarse grained.

148-162cm: Mostly shell fragments and some whole shell.

162-169cm: Siliceous.

169-172cm: 50/50 mix of sand and shell fragments.

172cm: Gradational contact.

172-182cm: Silty mud, stiff; well developed horizontal laminae, 1-5mm thick.

182-360cm: Silty clay and clayey silt, siliceous.

182-304cm: Extensively laminated, 1-20mm thick, mostly horizontal, but some cross-lamination, some 2.5Y6/2.

209-210.5cm: Sand lens.

211cm: Discontinuity.

238cm: Discontinuity.

304-360cm: Laminated, laminae 0.5-5cm.

317-327cm: Small shell fragments.

330-360cm: 5 sandy intervals, 0.5cm thick.

Core Number: ABUQ 13B
Location: N 31° 18.85' E 30° 5.18'
Water Depth: ~3m

Abu Qir/Alexandria 2001

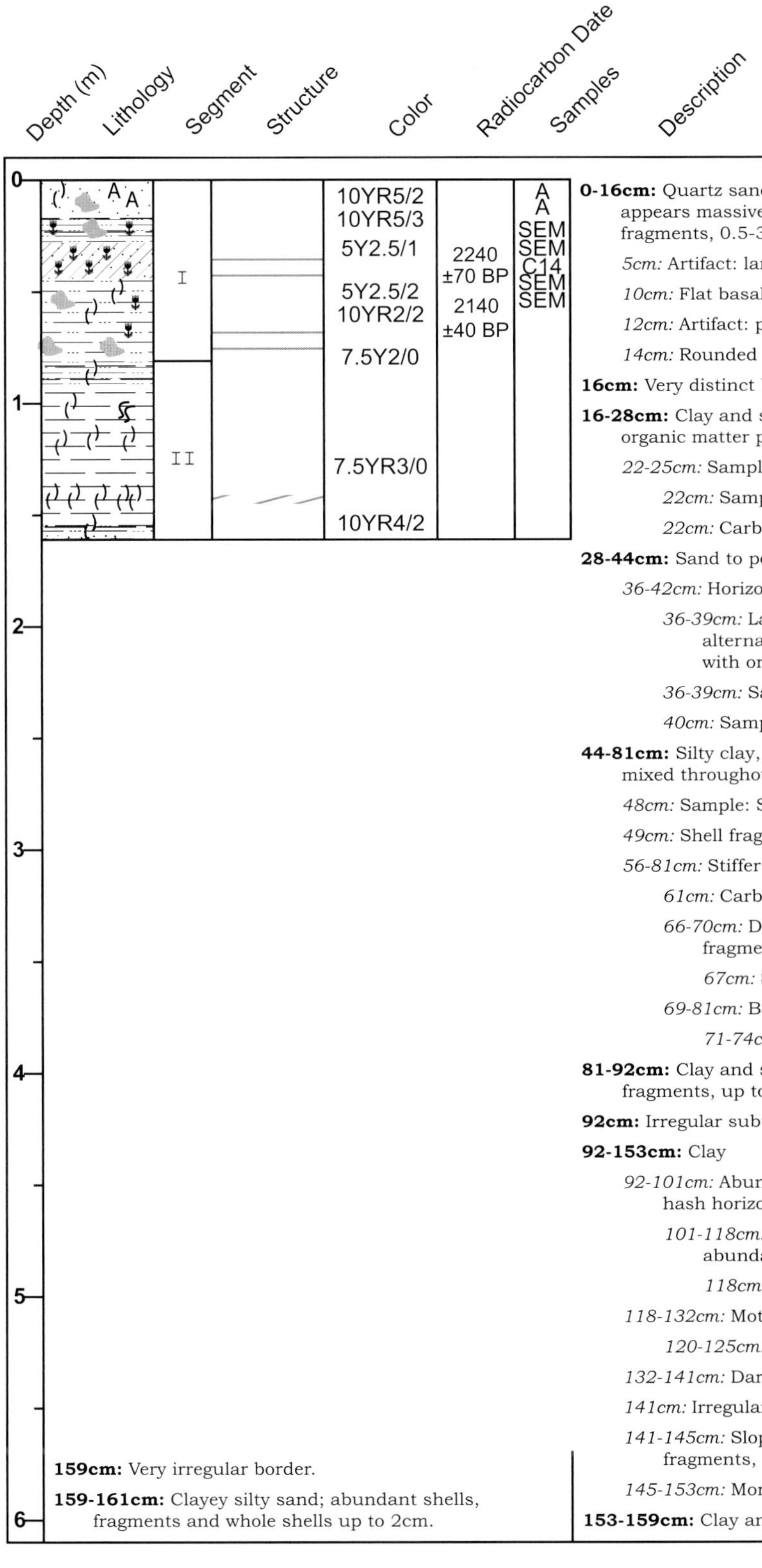

0-16cm: Quartz sand, fine grained, sub-angular to sub-rounded; sand appears massive; abundant shell, several varieties, complete and fragments, 0.5-3cm in size.

5cm: Artifact: large ceramic fragment; with kurkar fragment, 5x4x2cm.

10cm: Flat basalt rock fragment, polished, 2cm.

12cm: Artifact: possible shell.

14cm: Rounded shell fragment with heavy encrustation, 5cm.

16cm: Very distinct boundary.

16-28cm: Clay and silty clay, dense, compact, heavily mottled, micaceous; organic matter present.

22-25cm: Sample: SEM; organic matter.

22cm: Sample: SEM; decomposed, noncarbonized plant fragment.

22cm: Carbonate nodules, 3cm.

28-44cm: Sand to peaty sand, fine grained; organic rich; heavily mottled.

36-42cm: Horizontally bedded; organics up to 5cm.

36-39cm: Laminated organics with silty clay and brown sand; alternating horizontally bedded clays and brown sands mixed with organics.

36-39cm: Sample: C14, standard.

40cm: Sample: SEM.

44-81cm: Silty clay, dark, dense; abundant organic matter, noncarbonized, mixed throughout; less micaceous.

48cm: Sample: SEM; plant stem.

49cm: Shell fragments.

56-81cm: Stiffer and denser.

61cm: Carbonate nodules, up to 1cm.

66-70cm: Darker banded laminations with nodules and shell fragments.

67cm: Snail shell fragment, 0.5cm.

69-81cm: Banded laminations

71-74cm: Nodular zone.

81-92cm: Clay and silty clay, dark, dense, plastic, micaceous; shell fragments, up to 0.5cm.; minor woody organic matter.

92cm: Irregular subtle boundary.

92-153cm: Clay

92-101cm: Abundant shell and shell fragments, no whole shells; shell hash horizon with clay matrix; brown sand filled tubules.

101-118cm: Homogeneous, mottled, dark, dense, less micaceous; abundant shell and shell fragments.

118cm: Shell fragment, 3cm.

118-132cm: Mottled, dark, plastic.

120-125cm: Banded layer of broken shell fragments.

132-141cm: Dark, plastic, homogeneous.

141cm: Irregular boundary.

141-145cm: Sloping strata; very dense layer of shells and shell fragments, 1mm-3cm, matrix of plastic clay with very minor mica.

145-153cm: More homogenous; very few shells.

153-159cm: Clay and silty clay, dark, plastic, minor mica.

159cm: Very irregular border.

159-161cm: Clayey silty sand; abundant shells, fragments and whole shells up to 2cm.

Core Number: ABUQ 13C
Location: N 31° 18.85' E 30° 5.18'
Water Depth: ~3m

Abu Qir/Alexandria 2001

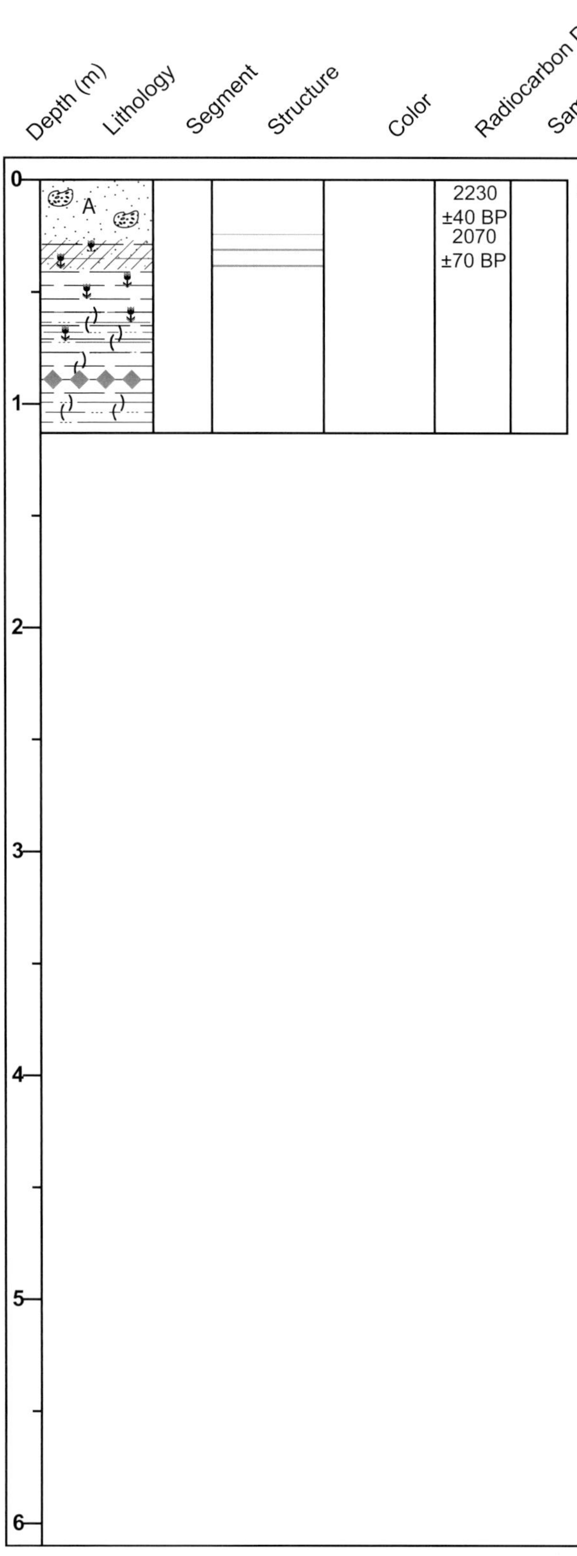

0-113cm: Hydrogen sulfide smell.

0-27cm: Quartz sand, fine grained, unconsolidated; sub-angular to sub-rounded grains; abundant kurkar, up to 9x6x4cm.

9cm: Artifact: ceramic.

11-15cm: Clay inclusion.

27cm: Irregular boundary.

27-32cm: Sandy peaty clay, dark, compact; minor organic material.

32cm: Steeply dipping boundary with discontinuity.

32-40cm: Peat, compact, laminated; sand and clay matrix; organic matter up to 3cm, woody, noncarbonized.

32-36cm: Sample: C14, standard.

33cm: Sample: SEM.

40cm: Irregular boundary.

40-56cm: Clay, very compact, dense, organic rich.

47cm: Sample: SEM.

56-81cm: Clay and silty clay, dark, very compact, dense, micaceous; organic matter up to 4cm; mottled.

58cm: Large shell fragment.

59cm: Sample: SEM.

65cm: Sample: SEM; large organic.

69cm: Shell fragment.

81-96cm: Clay, dark, compact, minor mica, some shell fragments.

91cm: Nodular horizon, nodules up to 1cm in size.

93cm: Carbonate nodules.

96-113cm: Silty clay, heavily micaceous, compact.

96-100cm: Sandy and shelly; shell material up to 0.5cm; less organic matter, very rare mica.

Core Number: ABUQ 14A

Location: N 31° 18.86' E 30° 5.35'

Water Depth: ~4m

Abu Qir/Alexandria 2001

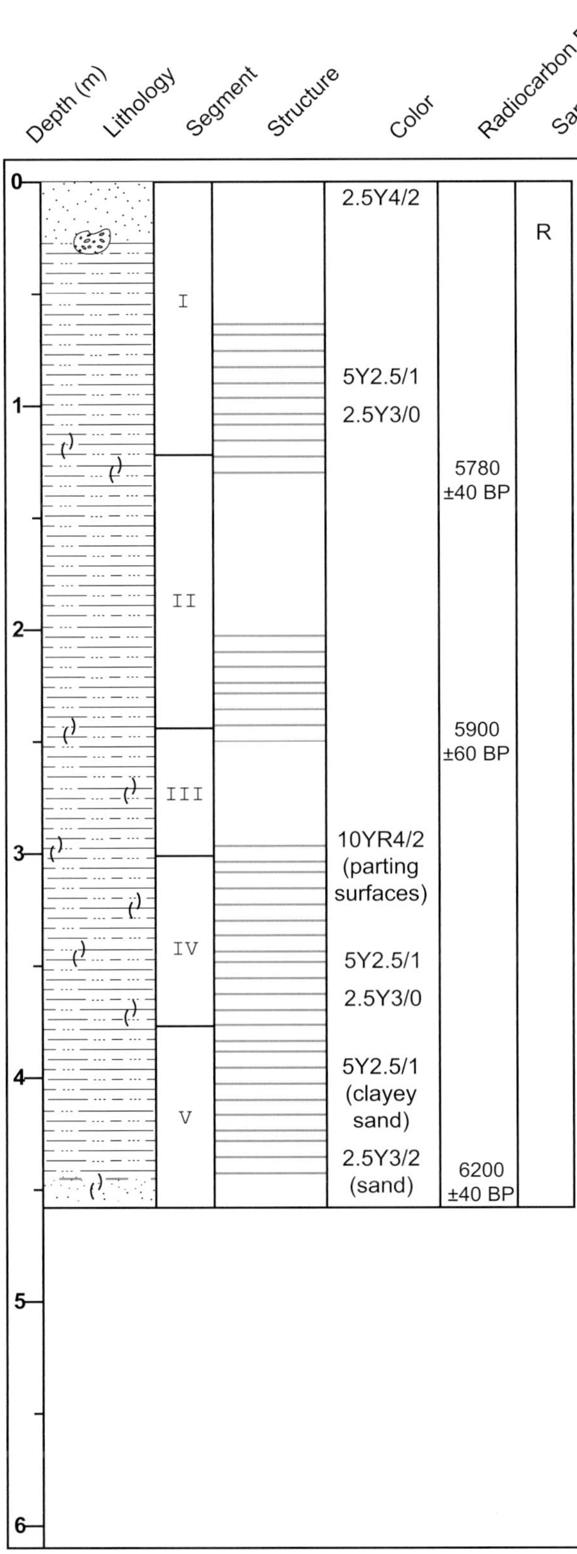

0-26cm: Sand, medium grained, 50/50 mix of abundant shell material and quartz.

16-26cm: Sample: rock.

26cm: Kurkar, 8x8x4cm.

26-444cm: Silty clay, siliceous, micaceous

26-244cm: Moderately soft.

60-122cm: Faint laminations, bedding on broken surfaces.

128-135: A few shell fragments.

200-244cm: Faint laminations.

244-377cm: Stiff, but sticky; scattered small shell fragments.

301-377cm: Faint laminations, ~5cm thick.

340-377cm: Distinctly colored (10YR4/2) parting surfaces, ~0.5cm thick. Microscopic inspection indicates some slight movement and water penetration (mineralization).

377-444cm: Faintly laminated, 0.5-5cm.

444-450cm: Clayey sand.

450-455cm: Sand, medium grained, clayey, mostly quartz but 25% shell.

455-458cm: Empty, no recovery.

Core Number: ABUQ 15A
Location: N 31° 18.87' E 30° 5.28'
Water Depth: ~3.5m

Abu Qir/Alexandria 2001

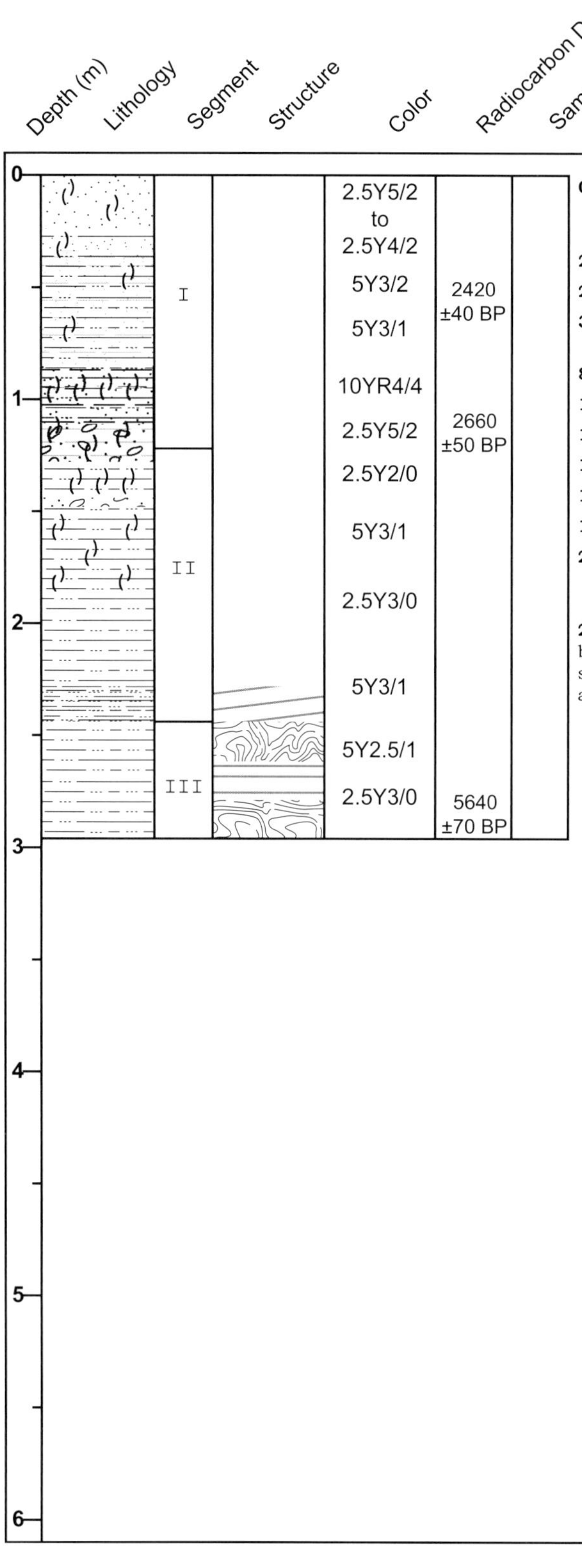

0-24cm: Sand, mostly medium grained, some macro shell fragments; sand ~60% quartz, 38% small shell fragments, 2% dark mineral; some small (<0.5cm) sandstone lithoclasts.

24-28cm: Silty clay.

28-34cm: Clay sand, fine grained; quartz and shell fragments.

34-85cm: Silty clay, with some very fine sandy clay; shell fragments common, some large; some subtle shell layers.

85-111cm: Silty clay and fine sandy clay with ~5 distinct shell layers.

111-128cm: Shell hash, variety of shell sizes; interval slightly muddy.

128-144cm: Shelly, silty clay, shells tend to occur in layers.

144-148cm: Shell hash.

148-161cm: Shelly silty clay.

161-228cm: Silty clay, siliceous, micaceous, only a few shell fragments.

228-244cm: Silty clay and clayey silt, siliceous, particularly micaceous, rather stiff; very well laminated, laminae from 0.5-10mm, generally parallel at 5-10∞.

244-296cm: Silty clay, siliceous, micaceous, rather stiff; faint laminations, but tends to break into play pieces. Distinctly colored parting surfaces, some indication of movement and enhanced flow of ground water; area around parting surfaces in 2.5Y4/2, very fine sand along parting surfaces.

Core Number: ABUQ 16A
Location: N 31° 18.95' E 30° 5.18'
Water Depth: 2.1m

Abu Qir/Alexandria 2001

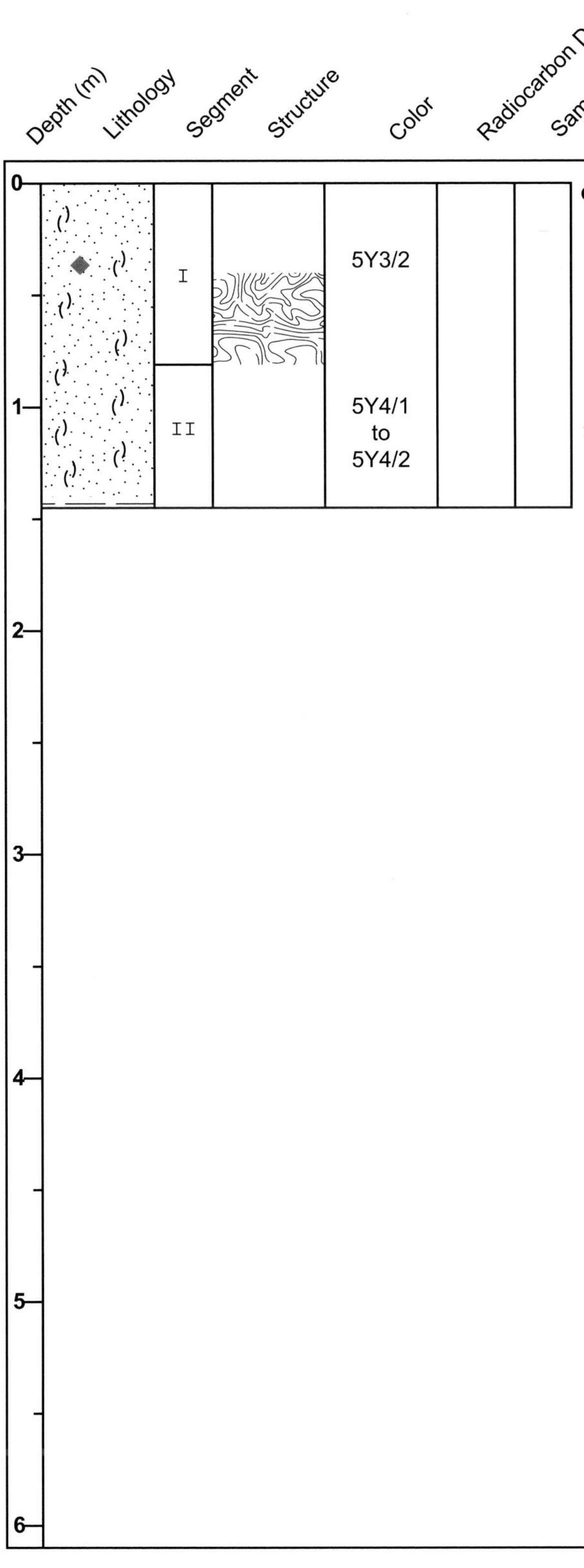

0-140cm: Sand, fine grained; mottled appearance.

0-81cm: Siliceous, 20% small shell fragments, ~2-3% dark minerals, slightly clayey in places; scattered shell and larger shell fragments.

40cm: A few concretions.

40-81cm: Distinctive knarly appearance, possibly result of liquefaction(?).

81-140cm: ~35% shell fragments, ~2% dark minerals; dense and compact.

140-145cm: Muddy.

Core Number: ABUQ 16B
Location: N 31° 18.95' E 30° 5.18'
Water Depth: 2.1m

Abu Qir/Alexandria 2001

Depth (m)	Lithology	Segment	Structure	Color	Radiocarbon Date	Samples
0	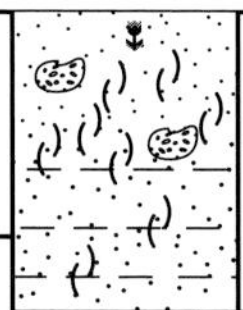	I		2.5Y5/2 10YR4/2 10YR4/1 10YR3/1 5YR4/1	6920 ±60 BP	C14
1						
2						
3						
4						
5						
6						

Description

0-9cm: Sand, loose, clean, very fine grained, sub-angular to rounded quartz grains.

3-4.5cm: Sample: C14, AMS.

9cm: Kurkar piece., 4cm in size.

9-26cm: Sand, loosely compact, fine to very fine grained, sub-angular to sub-rounded quartz, extremely heavy shell component; shells 2mm-2cm, some complete; wide variety of shells and modern roots scattered throughout.

26cm: Kurkar, heavily encrusted with shell, modern roots, 8x9x4cm.

26-34cm: Silty sand, dark, fine grained; shell material up to 3cm.

34cm: Gradational boundary.

34-67cm: Silty sand, fine grained, mottled with silt concentrations; minor to moderate amounts of shell, 2m-1cm; compact.

52-67cm: Partial recovery.

Core Number: ABUQ 17A

Location: N 31° 18.90' E 30° 5.29'

Water Depth: ~3.5m

Abu Qir/Alexandria 2001

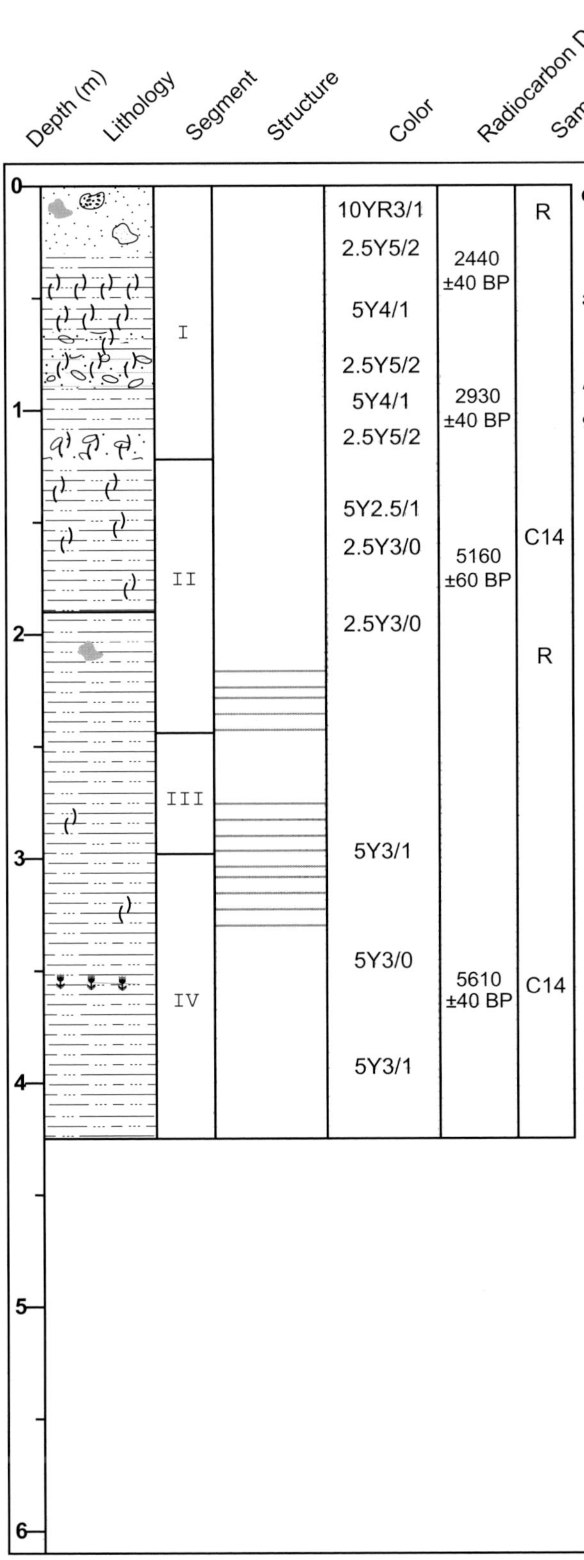

0-30cm: Sand, mostly medium grained; lithoclasts to 2cm common; 1 large kurkar fragment (5cm), others sandstone encrusted with algae; some macro shell fragments, small shell fragments common.

12cm: Sample: rock, black mineral present.

30-75cm: Silty clay, rather soft; shell fragments common.

45-48cm: Shell layer.

65-70cm: Shell hash.

75-90cm: Shell hash, clayey; many whole shells.

90-110cm: Silty clay; shell fragments common.

110-122cm: Shell hash.

122-190cm: Silty clay, rather stiff; tends to break in distinct plates.

122-130cm: Shelly.

130-190cm: Small shell fragments common.

153-159cm: Shelly.

157-158cm: Sample: C14, AMS, well preserved reed(?) material. Insufficient material for dating.

190cm: Discontinuity.

190-425cm: Silty clay, siliceous, micaceous.

210cm: Lithoclast, fine grained sandstone; contains plant fragments, some striations; 7x7.5x3cm; color 2.5Y5/2.

215-244cm: More clayey and stiffer; laminated, laminae 0.5-5cm at ~10∞.

244-298cm: Rather stiff; a number of parting surfaces stained 2.5Y3/2.

244-270cm: Faint laminae 3-5cm, horizontal.

298-425cm: Rather stiff.

298-360cm: A few shell fragments; fairly well developed laminae, 1-4cm, horizontal.

350-360cm: Plant-rich layer.

355-359cm: Sample: C14, AMS, well preserved wood(?); good quantity.

360-425cm: A number of parting surfaces stained 2.5Y3/2.

410-425cm: Quite stiff.

Appendix 2: AMS dates from Aboukir Bay vibracores

Calibrated Dates from Table 5.1 (pg 90-91)

SI ID	*Lab.No.*	*Uncalibrated date*	*BPError*	*Calibrated date ranges (- = BC)* *1 standard deviation*		*2 standard deviations*	
0022UB3883	145896	2250	60	-390	-200	-410	-160
0022UB3883	145897	2330	40	-490	-260	-520	-230
0022UB3883	145898	2010	40	-50	55	-160	80
0022UB3883	145899	3130	40	-1450	-1320	-1500	-1310
0022UB3883	145900	2070	40	-170	-40	-200	20
0022UB3883	145901	6440	70	-5480	-5350	-5540	-5290
0022UB3883	145902	2270	40	-400	-230	-400	-200
0022UB3883	145903	5620	50	-4500	-4360	-4550	-4350
0022UB3883	145904	6750	60	-5710	-5620	-5750	-5550
0022UB3883	145905	2070	40	-170	-40	-200	20
0022UB3883	145906	2330	50	-510	-250	-750	-200
0022UB3883	145907	2360	40	-510	-380	-740	-370
1A I	169951	5750	40	-4680	-4540	-4710	-4490
1A III	169964	6240	110	-5320	-5050	-5500	-4900
1A VI	159415	6620	60	-5620	-5515	-5640	-5470
1A VI	159416	7030	50	-5990	-5870	-6010	-5790
3B II	159417	6310	60	-5350	-5210	-5470	-5070
3B III	159418	5820	40	-4730	-4610	-4790	-4550
3B III	159419	6640	40	-5620	-5540	-5640	-5490
4A I	169952	3430	60	-1880	-1640	-1900	-1530
4A II	159420	5520	50	-4450	-4330	-4460	-4260
5A II	159422	6150	60	-5210	-5030	-5300	-4930
5A IV	159423	5640	40	-4530	-4400	-4550	-4360
5B II	159822	5940	50	-4900	-4720	-4950	-4710
5B II	159823	5700	50	-4600	-4460	-4690	-4440
7A I	169953	2780	40	-1000	-850	-1020	-820
7A II	169954	3710	40	-2200	-2030	-2280	-1970
8B I	159821	6390	50	-5470	-5310	-5480	-5290
9A I	169955	540	40	1320	1430	1300	1450
9A IV	159424	4100	40	-2850	-2570	-2880	-2490
10A II	159425	6300	60	-5350	-5210	-5470	-5070
11A I	169971	1570	40	430	540	400	580
11A I	169956	3270	40	-1610	-1490	-1640	-1440
11A II	159426	5030	50	-3950	-3710	-3960	-3700
11A II	159427	6250	40	-5305	-5205	-5320	-5060
11A III	159428	6080	40	-5060	-4930	-5210	-4840
12A I	169965	5990	60	-4950	-4790	-5020	-4720
12A II	169966	5940	60	-4900	-4720	-4990	-4690
12A III	169967	6190	60	-5220	-5050	-5310	-4990
12A IV	159430	5700	40	-4590	-4460	-4690	-4450
13A II	159431	1980	40	-40	65	-90	130
13A II	169957	2320	40	-420	-250	-520	-210
13A IV	169958	3770	40	-2290	-2130	-2310	-2030
13B I	159825	2220	70	-380	-200	-410	-90
13B I	168608	2120	40	-200	-50	-360	-40
13C I	168609	2240	40	-390	-210	-400	-200
13C I	159824	2050	70	-170	20	-350	90
14A II	169959	5880	40	-4795	-4710	-4850	-4610
14A III	169968	6000	60	-4960	-4790	-5040	-4720
14A V	169960	6570	40	-5550	-5480	-5620	-5470
15A I	169961	2880	40	-1130	-1000	-1210	-920
15A I	169972	3100	60	-1440	-1290	-1500	-1210
15A III	169969	5740	70	-4690	-4500	-4780	-4440
16B I	159826	6880	60	-5840	-5710	-5900	-5640
17A I	169962	2860	40	-1120	-940	-1190	-910
17A I	169963	3340	40	-1690	-1530	-1740	-1520
17A II	169970	5240	70	-4230	-3970	-4270	-3940
17A IV	159433	5700	40	-4590	-4460	-4690	-4450

Dates calibrated with OxCal version 3.10, © C. Bronk Ramsay 2005
Calendric dates converted from the conventional radiocarbon dates listed in table 5.1 (page 90).

Bibliography

Abdel-Kader, A. 1982. Landsat Analysis of the Nile Delta, Egypt. MSc thesis, University of Delaware, Newark

Abdou, H.F., Samir, A.M. and Frihy, O.E. 1991. Distribution of benthic foraminifera on the continental shelf off the Nile Delta. *Neues Jahrbuch für Geologie und Paläontologie*, Monatshefte, 1, 1–11

Adamson, D., Gasse, F., Street, F. and Williams, M. 1980. Late Quaternary history of the Nile. *Nature* 288, 50–5

Albani, A.D. and Serandrei Barbero, R. 1990. I foraminiferi della Laguna e del Golfo di Venezia. *Memorie di Scienze Geologiche* 42, 271–341

Alexandersson, T. 1990. Holocene shelf sedimentation of the western Egyptian shelf. *Rapports et Procès Verbaux de Réunions, Commission Internationale pour l'Exploration Scientifique de la Mer Méditerranée* 31(2), 108

Altersuch, J. 1979. The ecology and distribution of the littoral ostracods of Cyprus. *Journal of Natural History* 13, 135–60

Arbouille, D. and Stanley, J.-D. 1991. Late Quaternary evolution of the Burullus lagoon region, north-central Nile delta, Egypt. *Marine Geology* 99, 45–66

Arbulla, D., Pugliese, N. and Colonello, B. 2001. Ostracodi peritidali dell'area di S. Bartolomeo (Muggia, Italia). *Hydrores* 20, 78–84

Arrowsmith, A. 1802. *Plan of the operations of the British and Ottoman Forces in Egypt*. London (map, 1 sheet)

Arrowsmith, A. 1807. *A map of Lower Egypt from various surveys communicated by Major Bryce and other officers*. London (map, 1 sheet)

Attia, M.I. 1954. *Deposits in the Nile Valley and the Delta*. Geological Survey Publication 65. Cairo, Geological Survey of Egypt

Ball, J. 1939. *Contributions to the geography of Egypt*. Cairo, Survey and Mines Department

Ball, J. 1942. *Egypt in the classical geographers.* Cairo, Survey of Egypt, Government Press

Barbeito-Gonzalez, P.J. 1971. Die Ostracoden des Küstenbereiches von Naxos (Griechenland) und ihre Lebensbereiche. *Mitteilungen Hamburg Zoologischen Museum und Institut* 67, 255–326

Basso, D. and Corselli, C. 2002. Community versus biocoenosis in multivariate analysis of benthic molluscan thanatocoenoses. *Rivista Italiana di Paleontologia e Stratigrafia* 108(1), 153–72

Bates, C.C. 1953. Rational theory of delta formation. *The American Association of Petroleum Geologists Bulletin* 37, 2119–62

Bell, B. 1970. The oldest records of Nile floods. *The Geographical Journal* 136, 569–73

Bell, B. 1975. Climate and history of Egypt, the Middle Kingdom. *American Journal of Archaeology* 79, 223–69

Bernand, A. 1970. *Le delta égyptien d'après les textes grecs*, Cairo, Institut Français d'Archéologie Orientale du Caire

Bernasconi, M.P. and Stanley, J.-D. 1994. Molluscan biofacies and their environmental implications, Nile delta lagoons, Egypt. *Journal of Coastal Research* 10(2), 440–65

Bernasconi, M.P. and Stanley, J.-D. 1997. Molluscan biofacies, their distributions and current erosion on the Nile Delta shelf. *Journal of Coastal Research* 13(4), 1201–12

Bernasconi, M.P., Stanley, J.-D. and Di Geronimo, I. 1991. Molluscan fauna and paleobathymetry of Holocene sequences in the northeastern Nile Delta, Egypt. *Marine Geology* 99, 29–43

Blodget, H.W., Taylor, P.T. and Roark, J.H. 1991. Shoreline changes along the Rosetta-Nile promontory: monitoring with satellite observations. *Marine Geology* 99, 67–77

Bonaduce, G., Ciampo, G. and Masoli, M. 1975. Distribution of Ostracoda in the Adriatic Sea. *Pubblicazioni della Stazione Zoologica di Napoli* 40 (suppl.): 1–304

Bonaduce, G., Masoli, M. and Pugliese, N. 1988. Remarks on the benthic Ostracoda of the Tunisian shelf. In T. Hanai, N. Ikeya and K. Ishizaki (eds.), *Evolutionary biology of Ostracoda.* Proceedings of the 9th International Symposium on Ostracoda, Shizuoka. Amsterdam, Elsevier, 449–66

Bonaduce, G. and Pugliese, N. 1975. Ostracoda from Libya. *Pubblicazioni della Stazione Zoologica di Napoli* 39, 129–35

Breman, E. 1975. The distribution of ostracods in the bottom sediments of the Adriatic Sea. Unpublished thesis, Vrije Universiteit te Amsterdam, Academisch Preefscrift, 1–165

Broussard, M.L. (ed.), 1975. *Deltas, models for exploration*. Houston, TX, Houston Geological Society

Butzer, K.W. 1960. On the Pleistocene shorelines of Arabs' Gulf, Egypt. *Journal of Geology* 68, 626–37

Butzer, K.W. 1976. *Early hydraulic civilization in Egypt*. Chicago, University of ChicagoPress

Butzer, K.W. and Hansen, C.L. 1968. *Desert and river in Nubia: geomorphology and prehistoric environments at the Aswan Reservoir*. Madison, WI, University of Wisconsin Press

Carbonel, P. and Pujos, M. 1982. Comportement des microfaunes benthiques en milieu lagunaire. In *1er Congrès National en Sciences de la Terre*. Tunis, Université de Tunis, 127–39

Chen, Z., Warne, A.G. and Stanley, D.J. 1992. Late Quaternary evolution of the northwest Nile delta between Rosetta and Alexandria, Egypt. *Journal of Coastal Research* 8(3), 527–61

Cimerman, F. and Langer, M.R. 1991. Mediterranean foraminifera. *Slovenska Akademija Znanosti in Umetnosti. Opera Academia Scientiarium et Artium Slovenica*, Classis 4, Historia Naturalis, 30, 1-118

Clarke, K.R. and Warwich, R.M. 2001. *Change in marine communities: an approach to statistical analysis and interpretation* (second edition, in 2 vols.). Plymouth, PRIMER-E Ltd

Coleman, J.M. 1982. *Deltas: processes of deposition and models for exploration* (second edition). Boston, MA, International Human Resources Development

Coleman, J.M. 1988. Dynamic changes and processes in the Mississippi River delta. *Geological Society of America Bulletin* 100, 999–1015

Coleman, J.M., Roberts, H.H., Murray, S.P. and Salama, M. 1980. The morphology and dynamic sedimentology of the eastern Nile delta shelf. *Marine Geology* 41, 325–39

Coleman, J.M. and Wright, L.D. 1975. Modern river deltas, variability of processes and sand bodies. In M.L. Broussard (ed.), *Deltas: models for exploration*. Houston, Houston Geological Society, 99–149.

Constanty, H. 2002. Héracléion, les trésors de la ville engloutie. *GEO* 283, 148–58

Coulson, W.D.E. and Leonard, A. 1979. A preliminary survey of the Naukratis region in the western Nile delta. *Journal of Field Archaeology* 6, 151–68

De Cosson, A. 1935. *Mareotis*. London, Morrison and Gibb Ltd

Deibis, S. 1982. Aboukir Bay, a potential gas province area offshore Mediterranean, Egypt. *WEPCO, 6th Exploration Seminar of EGCP, Cairo, Egypt*, 50–65

Di Geronimo, I. and Robba, E. 1976. Metodologie qualitative e quantitative per lo studio delle biocenosi e delle paleocomunità bentoniche marine. *C.N.R. Gruppo Paleobenthos, Rapporto di Lavoro no. 1*, 1-35

Diodorus Siculus (*c.* 80–20 BC). *Library of History, Book I*, trans. C.H. Oldfather. London, Loeb Classical Library.

Du Bois-Aymé, M. 1813. Mémoire sur les anciennes branches du Nil et ses embouchures dans la mer. **In** *Description de l'Égypte, Antiquités, Mémoires*, I, 277–90

Eastman Kodak Company 2000 Health Imaging Data Sheet TI1816, Kodak Ektascan M Film/6515. Rochester, NY

El Askary, M.A. and Frihy, O.E. 1986. Depositional phases of Rosetta and Damietta promontories on the Nile delta coast. *Journal of African Earth Sciences* 5, 627–33

El Bouseily, A.M. and Frihy, O.E. 1984. Textural and mineralogical evidence denoting the position of the mouth of the old Canopic Nile branch on the Mediterranean coast, Egypt. *Journal of African Earth Sciences* 2, 103–7

El Din, S.S.H. 1977. Effect of the High Aswan Dam on the Nile flood on estuarine and coastal circulation along the Mediterranean Egyptian coast. *Limnology and Oceanography* 22, 194–201

El Fattah, T.A. and Frihy, O.E. 1988. Magnetic indications of the position of the mouth of the old Canopic branch of the northwestern Nile delta of Egypt. *Journal of Coastal Research* 4, 483–8

El-Fayoumy, I.F., El Shazli, M.M. and Hammad, F.A. 1975. *Geomorphology of the coastal area between Aboukir and Rasheed (north west of the Nile delta, E.A.R.).Bulletin of the Faculty of Science 2*. Cairo, Mansoura University Press, 135–47

El Fishawi, N.M. and El Askary, M.A. 1981. Characteristic features of coastal sand dunes along Burullus–Gamasa stretch, Egypt. *Acta Mineralogica-Petrographica, Szeged* 25, 63–76

El Fishawi, N.M. and Fanos, A.M. 1989. Prediction of sea level rise by 2100, Nile delta coast. *INQUA Commission on Quaternary Shorelines Newsletter* 11, 43–7

El Ramly, I.M. 1968. Quaternary shoreline changes relative to paleoclimates and their emphasis on groundwater possibilities of the El Amiriya/El Alamein district, Western Desert, Mediterranean littoral, U.A.R. *Bulletin de l'Institut d'Égypte* 49, 163–6

El Ramly, I.M. 1971. Shoreline changes during the Quaternary in the Western Desert Mediterranean coastal region (Alexandria–Sallum), U.A.R. *Quaternaria* 15, 285–92

El-Sayed, M.K. 1979. The inner shelf off the Nile delta: sediment types and depositional environments. *Oceanologica Acta* 2, 249–52

El-Sayed, M.K. 1996. Rising sea level and subsidence of the northern Nile delta: a case study. In J.D. Milliman and B.U. Haq (eds.), *Sea level rise and coastal subsidence.* Dordrecht, Kluwer Academic Publishers, 215–33

El Shazly, E.M., Abdel-Hady, M.A., Ghawaby, M.A., El Kassas, I.A., Khawasik, S.M., El Shazly, M.M. and Sanad, S. 1975. *Geologic interpretation of Landsat satellite images for West Nile Delta area, Egypt.* Cairo, The Remote Sensing Research Project, Academy of Scientific Research and Technology.

El-Wakeel, S.K., Abdou, H.F. and Wahby, S.D. 1970. Foraminifera from bottom sediments of Lake Maryut and Lake Manzalah, Egypt. *Bulletin of the Institute of Oceanography and Fisheries* 1, 429–48

El-Wakeel, S.K. and El Sayed, M. 1978. The texture, mineralogy and chemistry of bottom sediments and beach sands from the Alexandria region, Egypt. *Marine Geology* 27, 137–60

Elliott, T. 1986. Deltas. In H.G. Reading (ed.), *Sedimentary environments and facies* (second edition). Oxford, Blackwell Scientific Publications, 113–54

Fairbanks, R.G. 1989. A 17,000-year glacio-eustatic sea level record: influence of glacial melting rates on the Younger Dryas event and deep-ocean circulation. *Nature* 342, 637–42

Fanos, A.M. 1986. Statistical analysis of longshore current data along the Nile delta coast. *Water Science Journal* 1, 45–55

Fiorini, F. and Vaiani, S.C. 2001. Benthic foraminifers and transgressive–regressive cycles in the Late Quaternary subsurface sediments of the Po Plain near Ravenna (northern Italy). *Bollettino della Società Paleontologica Italiana* 40(3), 357–403

Fourtau, R. 1915. Contribution à l'étude des dépôts nilotiques. *Mémoires de l'Institut Egyptien* 8, 57–94

Frihy, O.E. 1988. Nile delta shoreline changes: aerial photographic study of a 28-year period. *Journal of Coastal Research* 4(40), 597–606

Frihy, O.E. 1992a. Quaternary evolution of Aboukir Bay, Egypt. *GeoJournal* 27(4), 339–44

Frihy, O.E. 1992b. Sea level rise and shoreline retreat of the Nile delta promontories, Egypt. *Natural Hazards* 5, 65–81

Frihy, O.E., El Fishawi, M.M. and El Askary, M.A. 1988. Geomorphological features of the Nile delta coastal plain: a review. *Acta Adriatica* 29, 51–65

Frihy, O.E. and Komar, P.O. 1991. Patterns of beach-sand sorting and shoreline erosion on the Nile delta. *Journal of Sedimentary Petrology* 61, 544–50

Frihy, O.E., Moussa, A.A. and Stanley, J.-D. 1994. Aboukir Bay, a sediment sink off the northwestern Nile delta, Egypt. *Marine Geology* 121, 199–211

Fujiwara, O., Masuda, F., Sakai, T., Irizuki, T. and Fuse, K. 2000. Tsunami deposits in Holocene bay mud in southern Kanto region, Pacific coast of central Japan. *Sedimentary Geology* 135, 219–30

Galloway, W.E. 1975. Process framework for describing the morphologic and stratigraphic evolution of deltaic depositional systems. In M.I. Broussard (ed.), *Deltas, models for exploration*. Houston, Houston Geological Society, 87–98

Goddio, F. 2007. *The Topography and Excavation of Heracleion-Thonis and East Canopus (1996–2006).* (Underwater Archaeology in the Canopic Region in Egypt). Oxford, OCMA

Guidoboni, E. 1994. *Catalogue of ancient earthquakes in the Mediterranean area up to the 10th century.* Bologna, Istituto Nazionale di Geofisica

Halim, Y. and Gerges, S.K. 1981. Coastal lakes of the Nile delta. Lake Manzala. *Symposium on Coastal Lagoons, UNESCO, Technical Papers in Marine Science* 33, 135–72

Hamid, S. 1984. Fourier analysis of Nile flood levels. *Geophysical Research Letters* 11,843–58

Hassan, F.A. 1981. Historical Nile floods and their implications for climatic change. *Science* 212, 1142–5

Hassouba, A.M.B.H. 1980. Quaternary sediments from the coastal plain of northwestern Egypt from Alexandria to El Omayid. Unpublished PhD thesis, Imperial College London

Hassouba, A.M.B.H., 1995. Quaternary sediments from the coastal plain of northwestern Egypt (from Alexandria to El Omayid). *Carbonates and Evaporites* 10(1), 8–44

Herodotus, *The History*, trans. David Grene, 1987. Chicago, University of Chicago Press

Ho, C. and Coleman, J.M. 1969. Consolidation and cementation of recent sediments in the Atchafalaya basin. *Geological Society of America Bulletin* 80(2), 183–92

Hoffman, M.A. 1979. *Egypt before the Pharaohs*. New York, Knopf

Holeman, J.N. 1968. The sediment yield of major rivers of the world. *Water Resources Research* 4, 737–47

Hottinger, L., Halicz, E. and Reiss, Z. 1993. Recent foraminiferida from the Gulf of Aqaba, Red Sea. *Slovenska Akademija Znanosti in Umetnosti. Opera Academia Scientiarium et Artium Slovenica.*, Classis 4, Historia Naturalis, 33

Hurst, E.H. 1931–66. *The Nile Basin*. Cairo, Physical and Nile Control Department, Ministry of Public Works (9 vols.)

Hussein, I.M. and Abd-Allah, A.M.A. 2001. Tectonic evolution of the northeastern part of the African continental margin, Egypt. *Journal of African Earth Sciences* 33(1), 49–68

Inman, D.L. and Jenkins, S.A. 1984. The Nile littoral cell and man's impact on the coastal zone of the southeastern Mediterranean. *Scripps Institution of Oceanography, Reference Series* 84–31, 1–43

Ippoliti, R. 1993–4. Gli ostracodi della laguna Idku e della baia di Aboukir (Delta del Nilo, Egitto) quali indicatori ambientali. Unpublished thesis, University of Trieste

IWACO, Consultants for Ground Water and Environment 1989. Landsat Thematic Mapper for hydrogeological mapping in Egypt. *Report bcrs 89–28, Final report CO-1.7, Development and Management of Groundwater Resources in the Nile Valley and Delta Project.* Rotterdam

Jacotin, E. 1826. *Atlas géographique: description de l'Égypte*. Paris, Panokouke

Jacotin, P.M. 1818. *Carte topographique de l'Égypte et de plusieurs parties des pays limitrophes*. Paris

Kebeasy, R.M. 1990. Seismicity. In R. Said (ed.), *The Geology of Egypt*. Rotterdam, A.A. Balkema, 51–9

Kerambrun, P. (ed.) 1986. *Coastal lagoons of the southern Mediterranean coast (Algeria, Egypt, Libya, Morocco, Tunisia): description and bibliography.* UNESCO Reports in Marine Science 34. Paris, UNESCO

Kulyk, V.A. 1987. Holocene foraminifera of the eastern Nile delta, Egypt. MSc thesis, George Washington University, Washington, DC

Langer, M. 1988. Recent epiphytic foraminifera from Vulcano (Mediterranean Sea). *Revue de Paléobiologie*, Special Issue 2, 827–32

Le Calvez, J. and Le Calvez, Y. 1958. Repartition des foraminifères dans la Baie de Villefranche. *Annales de l'Institut Océanographique* 35, 159–234

Lévêque, P. 1994. *The birth of Greece*. New York, Harry N. Abrams

Levy, A., Mathieu, R., Poignant, A. and Rousset-Moulinier M. 1992 Foraminfers à arrangement quinqueloculin et triloculin (Miliolacea de Méditerranée). *Revue de Paléobiologie* 11(1), 111-35

Loizeau, J.-L. and Stanley, J.-D. 1993. Petrological-statistical approach to interpret recent and sub-recent lagoon subfacies, Idku, Nile delta of Egypt. *Marine Geology* 111, 55–81

Maestro, A., Barnolas, A., Somoza, L., Lowrie, A. and Lawton, T. 2002. Geometry and structure associated to gas-charged sediments and recent growth faults in the Ebro Delta (Spain). *Marine Geology* 186, 351–68

Manohar, M. 1981. Coastal processes at the Nile delta coast. *Shore and Beaches* 49, 8–15

Mansouri, R. and Carbonel, P. 1981. Les déplacements ostracodes et leur relation avec les courants dans la lagune de Ghar el Melah. *Bulletin Société des Sciences Naturelles, Tunisie* 16, 55–6

Marocco, R., Melis, R., Montenegro, M.E., Pugliese, N., Vio, E. and Lenardon, G. 1996. Holocene evolution of the Caorle barrier-lagoon (northern Adriatic Sea, Italy). *Rivista Italiana di Paleontologia e Stratigrafia* 102(3), 385–96

Milliman, J.D. and Haq, B.U. (eds.) 1996. *Sea level rise and coastal subsidence.* Dordrecht, Kluwer Academic Publishers

Moncharmont-Zei, M., 1968. I foraminiferi di alcuni campioni di fondo prelevati lungo la costa di Beirut (Libano). *Bollettino della Società dei Naturalisti in Napoli* 77, 3–34.

Montenegro, M.E. and Pugliese, N. 1995. Ostracodi della laguna di Orbetello: tolleranza e opportunismo. *Atti del Museo di*

Geologia e Paleontologia di Monfalcone, Quaderno Spec. 3, 71–80

Montenegro, M.E. and Pugliese, N. 1996. Autecological remarks on the ostracod distribution in the Marano and Grado lagoons (Northern Adriatic Sea, Italy). *Bollettino della Società Paleontologica Italiana*, Spec. Vol. 3, 123–32

Montenegro, M.E., Pugliese, N. and Bonaduce, G. 1998. Shelf ostracods distribution in the Italian seas. In *'What about Ostracoda?' Actes 3e Congrès Européen des Ostracodologistes, Paris-Bierville, France, 1996. Bulletin Centre Recherches Elf Exploration Production (Pau) Mémoire* 20, 91–101.

Morgan, J.P. 1970. *Deltaic sedimentation, modern and ancient.* SEPM Special Publication 15. Tulsa, Society of Economic Paleontologists and Mineralogists

Morgan, J.P., Coleman, J.M. and Gagliano, S.M. 1963. *Mudlumps at the mouth of South Pass, Mississippi River: sedimentology, paleontology, structure, origin, and relation to deltaic processes.* Coastal Studies Institute Series 10. Baton Rouge, LA, Louisiana State University

Mörner, N.-A. 1971. Eustatic changes during the last 20,000 years and a method of separating the isostatic and eustatic factors in an uplifted area. *Paleogeography, Paleoclimatology, Paleoecology* 9, 153–81

Mostafavi, N. 1983. Kroemmelbeinia n.g., eine neue Ostracoden-Gattung aus dem marinen Oberpliozän der Insel Kos (Griechenland). *Paläontologische Zeitschrift* 57(1/2), 2 Abb., 69–74

Müller, G.W. 1894. Die Ostracoden des Golfes von Neapel und der angrezenden Meeresabschnitte. *Fauna und Flora des Golfes von Neapel* 21, Monographie, i-viii,1–104

Munsell Color 1975. *Munsell soil color charts.* Baltimore, MD, Kollmorgen Corp.

Murray, S.P., Coleman, J.M. and Roberts, H.H. 1981. Accelerated currents and sediment transport off the Damietta Nile promontory. *Nature* 293(5827), 51–4

Nafaa, M.E. and Frihy, O.E. 1993. Beach and nearshore features along the dissipative coastline of the Nile delta, Egypt. *Journal of Coastal Research* 9, 423–33

Nafas, M.G., Fanos, A.M. and Elganainy, M.A. 1991. Characteristics of waves off the Mediterranean coast of Egypt. *Journal of Coastal Research* 7(3), 665–76

Nur, A. 2000. Earthquakes and archeology in ancient Alexandria (abstract). *Eos, Transactions, American Geophysical Union* 81(48), F21

Pallary, P. 1911. Catalogue des mollusques du littoral méditerranéen de l'Égypte. *Mémoires de l'Institut Egyptien* 7(3), 69–200

Parker, F.L. 1958. Eastern Mediterranean foraminifera. *Swedish Deep-Sea Expedition* 8, 219-283.

Penland, S., Boyd, R. and Suter, J.R. 1988. Transgressive depositional systems of the Mississippi delta plain: a model for barrier shoreline and shelf sand development. *Journal of Sedimentary Petrology* 58(6), 932–49

Pérès, J.M. and Picard, J. 1964. Nouveau manuel de bionomie benthique de la Mer Méditerranée. *Receuil des Traveaux de la Station Marine d'Endoume* 31, 1–137

Petit-Maire, N. 1989. Interglacial environments in presently hyperarid Sahara: paleoclimatic implications. In M. Leinen and N. Sarnthein (eds.), *Paleoclimatology and paleometeorology: modern and past patterns of global atmospheric transport.* Boston, MA, Kluwer, 637–61

Pirazzoli, P.A. 1986. The early Byzantine tectonic paroxysm. *Zeitschrift für Geomorphologie*, N.F., supplement 62, 31–49

Pirazzoli, P.A. 1987. Sea level changes in the Mediterranean. In M.J. Tooley and I. Shennan (eds.), *Sea level changes.* Institute of British Geographers, Special Publication, Series 20. Oxford, Basil Blackwell, 152–81

Pirazolli, P.A. 1992. *World atlas of sea level changes.* Amsterdam, Elsevier Oceanography Series 58

Poem Group 1992. General circulation of the eastern Mediterranean. *Earth-Science Reviews* 32, 285–309

Popper, W. 1951. *The Cairo Nilometer.* Los Angeles, University of California Press

Prior, D.B. and Coleman, J.M. 1982. Active slides and flows in unconsolidated marine sediments on the slopes of the Mississipi delta. In S. Saxov and J.K. Nieuwenhuis (eds.), *Marine slides and other mass movements.* New York, Plenum Press, 21–49

Pugliese, N. and Stanley, J.-D. 1991. Ostracoda, depositional environments and late Quaternary evolution of the eastern Nile Delta, Egypt. *Il Quaternario, AIQUA* 4(2), 275–302

Quelennec, R.E. and Kruc, C.B. 1976. Nile suspended load and its importance for Nile delta morphology. In *UNDP/UNESCO Proceedings of the Seminar on Nile Delta Sedimentology.* Alexandria, Academy of Scientific Research and Technology, Egypt

Reinhardt, E.G., Patterson, R.T. and Schröeder-Adams, C.J. 1994. Geoarchaeology of the ancient harbour site of Caesarea Maritima, Israel: evidence from sedimentology and paleoecology of benthic foraminifera. *Journal of Foraminiferal Research* 24(1), 37–48

Riehl, H. and Meitin, J. 1979. Discharge of the Nile River: a barometer of short-period climatic variations. *Science* 206, 1178–9

Ross, D.A. and Uchupi, E. 1977. Structure and sedimentary history of southeastern Mediterranean Sea Nile Cone area. *Bulletin of the American Association of Petroleum Geologists* 61(6), 872–902

Said, R. 1981. *The geological evolution of the river Nile*. New York, Springer-Verlag

Said, R. (ed.) 1990. *The geology of Egypt*. Rotterdam, A.A. Balkema, 51–9

Said, R. 1993. *The river Nile: geology, hydrology and utilization*. New York, Pergamon Press

Samir, A.M. 2000. The response of benthic foraminifera and ostracods to various pollution sources: a study from two lagoons in Egypt. *Journal of Foraminiferal Research* 30(2), 83–98

Samir, A.M. and El Din, A.B. 2001. Benthic foraminiferal assemblages and morphological abnormalities as pollution proxies in two Egyptian bays. *Marine Micropaleontology* 41, 193–222

Schlumberger Corp., 1984. *Well Evaluation Conference; Egypt 1984, Middle East S.A.* Paris, Imprimerie Moderne du Lion

Sestini, G. 1989. Nile delta: a review of depositional environments and geological history. In M.G.K. Whateley and K.T. Pickering (eds.), *Deltas: sites and traps for fossil fuels*. London, Geological Society of London, Special Publications 41, 99–127

Sestini, G. 1992a. Implications of climatic changes for the Nile delta. In L. Jeftic, J.D. Milliman and G. Sestini (eds.), *Climatic change and the Mediterranean*. London, Edward Arnold, 535–601

Sestini, G. 1992b. Nile delta: a review of depositional environments and geological history. In M.G.K. Whateley and K.T. Pickering (eds.), *Deltas: sites and traps for fossil fuels*. London, Geological Society of London, Special Publications 41, 99–127

Sgarrella, F. and Moncharmont Zei, M. 1993. Benthic foraminifera of the Gulf of Naples (Italy): systematics and autoecology. *Bollettino della Società Paleontologica Italiana* 32(2), 145–264

Shafei, A. 1952. Lake Mareotis – its past history and its future development. *Bulletin de l'Institut Fouad I du Désert* 2(1), 71–101

Shahin, M. 1985. *Hydrology of the Nile Basin*. Amsterdam, Elsevier

Sharaf El Din, S.H. 1977. Effect of the Aswan High Dam on the Nile flood and on the estuarine and coastal circulation pattern along the Mediterranean Egyptian coast. *Limnology and Oceanography* 22(2), 194–207

Shata, A. 1955. An introductory note on the geology of the northern portion of the Western Desert of Egypt. *Bulletin de l'Institut du Désert d'Égypte* 5, 96–106

Shukri, N.M. 1950. The mineralogy of some Nile sediments. *Quaternary Journal of the Geological Society, London* 105, 511–34

Siliotti, A. 1998. *The discovery of Ancient Egypt*. Vercelli, Italy, Edizioni White Star

Slack, J.M., Kontroviz, M. and Stanley, J.-D. 1995. Ostracoda from Lake Manzala, Nile Delta, Egypt. In J. Riha (ed.), *Ostracoda and biostratigraphy*. Rotterdam, A.A. Balkema, 333–42

Smith, S.E. and Abdel-Kader, A. 1988. Coastal erosion along the Egyptian delta. *Journal of Coastal Research* 4, 244–55

Sneh, A., Weissbrod, T., Ehrlich, A., Horowitz, A., Moshkovitz, S. and Rosenfeld, A. 1986. Holocene evolution of the northeastern corner of the Nile Delta. *Quaternary Research* 26, 194–206

Soloviev, S.L., *et al.* 2000. *Tsunamis in the Mediterranean Sea 2000 BC–2000 AD*. Dordrecht, Kluwer Academic Publishers

Stanley, J.-D. 1988. Subsidence in the northeastern Nile delta: rapid rates, possible causes and consequences. *Science* 240, 497–500

Stanley, J.-D. 1990. Recent subsidence and northeast tilting of the Nile delta, Egypt. *Marine Geology* 94 , 147–54

Stanley, J.-D. 1996. Nile delta: extreme case of sediment entrapment on a delta plain and consequent coastal land loss. *Marine Geology* 129, 189–95

Stanley, J.-D. 1997. Mediterranean deltas: subsidence as a major control of relative sea level rise. *Bulletin de l'Institut Océanographique*, Monaco, Numéro special 18, 35–62

Stanley, J.-D. 2003. Nile delta margin: failed and fluidised deposits concentrated along distributary channels. *Géomorphologie* 4, 211–26

Stanley, J.-D. 2005a. Growth faults, a distinct carbonate-siliciclastic interface and recent coastal evolution, NW Nile delta, Egypt. *Journal of Coastal Research* 42, 309–18

Stanley, J.-D. 2005b. Submergence and burial of ancient coastal sites on the subsiding Nile delta, Egypt. *Méditerranée* 1(2), 65–73

Stanley, J.-D., Arad, V., Bartov, Y. and El Bedewy, F.M. 1993. *The Nile delta: bibliography of geological research.* Cairo and Jerusalem: Geological Survey of Egypt and Israel Geological Survey, Report GSI/22/93

Stanley, J.-D. and Bernasconi, M.P. 1998. Relict and palimpsest depositional patterns in the Nile shelf recorded by molluscan fauna. *Palaios* 13, 79–86

Stanley, J.-D. and Bernasconi, M.P. 2006. Holocene depositional patterns and evolution in Alexandria's Eastern Harbor, Egypt. *Journal of Coastal Research* 22, 283–97

Stanley, J.-D., Goddio, F., Jorstad, T. F. and Schnepp, G. 2004. Submergence of ancient Greek cities off Egypt's Nile delta – a cautionary tale. *GSA Today*. 14(1), 4–10

Stanley, J.-D., Goddio, F. and Schnepp, G. 2001. Nile flooding sank two ancient cities. *Nature* 412, 293–4

Stanley, J.-D. and Hamza, F.H. 1992. Terrigenous–carbonate sediment interface (late Quaternary) along the northwestern margin of the Nile delta, Egypt. *Journal of Coastal Research* 8, 153–71

Stanley, J.-D. and Jorstad, T.F. 2006. Buried Canopic channel identified near Egypt's Nile delta coast with radar (SRTM) imagery. *Geoarchaeology* 21, 503–14

Stanley, J.-D., McRea, J.E. Jr. and Waldron, J.C. 1996. *Nile delta drill core and sample databases for 1985–1994: Mediterranean Basin (MEDIBA) Program.* Smithsonian Contributions to the Marine Sciences 37. Washington, DC, Smithsonian Institution Press

Stanley, J.-D. and Warne, A.G. 1993a. Nile delta: recent geological evolution and human impact. *Science* 260, 628–34

Stanley, J.-D. and Warne, A.G. 1993b. Sea level and initiation of Predynastic culture in the Nile delta. *Nature* 363, 435–8

Stanley, J.-D. and Warne, A.G. 1994. Worldwide initiation of Holocene marine deltas: deceleration of sea level rise as principal factor. *Science* 265, 228–31

Stanley, J.-D. and Warne, A.G. 1998. Nile delta in its destruction phase. *Journal of Coastal Research* 14, 794–825

Stanley, J.-D. and Warne, A.G. 2006. Nile delta geography at the time of Herakleion and eastern Canopus. In J.-D. Stanley (ed.) Chapter 2 (this volume)

Stanley J.-D., Warne, A.G., Davis, H.R., Bernasconi, M.P. and Chen, Z. 1992. Late Quaternary evolution of the north-central Nile delta between Manzala and Burullus lagoons, Egypt. *National Geographic Research & Exploration* 8, 22–51

Stanley, J.-D., Warne, A.G. and Schnepp, G. 2004. Geoarchaeological interpretation of the Canopic, largest of the relict Nile distributaries, Egypt. *Journal of Coastal Research* 20, 920–30

Stiros, S.C. 2001. The Crete earthquake of 365 BC and possible seismic clustering during the fourth to sixth centuries BC in the eastern Mediterranean: a review of historical and archaeological data. *Journal of Structural Geology* 23, 545–62

Summerhayes, C.P., Sestini, G., Misdorp, R. and Marks, N. 1978. Nile delta: nature and evolution of continental shelf sediments. *Marine Geology* 27, 43–65

Takashimizu, Y. and Masuda, F. 2000. Depositional facies and sedimentary successions of earthquake-induced tsunami deposits in Upper Pleistocene incised valley fills, central Japan. *Sedimentary Geology* 135, 231–9

Toussoun, O. 1922. *Mémoires sur les anciennes branches du Nil époque ancienne*. Mémoire de l'Institut d'Égypte 4. Cairo, Institut Français d'Archéologie Orientale

Toussoun, O., 1926. *Mémoire sur l'histoire du Nil.* Mémoires de la Société Royale Archéologique d'Alexandrie 6. Cairo, Institut Français d'Archéologie Orientale

Toussoun, O. 1934. Les ruines sous-marines de la Baie d'Aboukir. *Bulletin de la Société Royale d'Archéologie, Alexandrie* 29, 342–52

UNDP/UNESCO 1977. *Proceedings of Seminar on Nile delta shore processes.* Alexandria, UNDP

UNDP/UNESCO 1978. *Coastal protection studies: project findings and recommendations.* Paris,UNDP/EGY/73/063

US Defense Mapping Agency 1977. Rosetta (Rashîd). Map Series P773, Sheet 5387–I, Scale 1:50,000, Washington, DC

Vismara Schilling, A. and Ferretti, L. 1987. Analisi semiquantitativa delle microfaune a foraminiferi e ostracodi nella laguna di S. Teodoro (Sardegna nord-orientale). Distribuzione delle associazioni in funzione del grado di confinamento. *Bollettino dell'Accademia Gioenia di Scienze Naturali* 20, 45–92

Waltham, T. 2002. Sinking cities. *Geology Today* 18, 95–100

Warne, A.G. and Stanley, J.-D. 1993. Late Quaternary evolution of the northwest Nile delta and adjacent coasts in the Alexandria region, Egypt. *Journal of Coastal Research* 9, 26–64

Warne, A.G. and Stanley, J.-D. 1995. Sea level change as critical factor in development of basin margin sequences: new evidence from late Quaternary record. *Journal of Coastal Research* Special Issue 17, 231–40

Waterbury, J. 1979. *Hydropolitics of the Nile Valley*. Syracuse, NY, Syracuse University Press

Wendorf, F., Schild, R., Said, R., Haynes, C.V., Gautier, A. and Kobusiewicz, M. 1976. The prehistory of the Egyptian Sahara. *Science* 193, 103–13

Wenke, R.J. 1991. The evolution of early Egyptian civilization: issues and evidence. *Journal of World Prehistory* 5(3), 279–329

Wheeler, R.L. 2002. Distinguishing seismic from nonseismic soft-sediment structures: criteria from seismic-hazard analysis. In F.R. Ettensohn, N. Rast and C.E. Brett (eds.), *Ancient Seismites*. Special paper 359. Boulder, CO, Geological Society of America, 1–11

Wouters, K. 1973. Quelques ostracodes du Tyrrhénien de Monastir (Tunisie). *Annales des Mines et de la Géologie* 26, 379–99.

Wright, L.D. and Coleman, J.M. 1973. Variations in morphology in major river deltas as functions of ocean waves and river discharge regimes. *The American Association of Petroleum Geologists Bulletin* 57, 370–98

Wright, L.D. and Coleman, J.M. 1974. Mississippi River mouth processes: effluent dynamics and morphologic development. *Journal of Geology* 82, 751–78

Yanko, V., Kronfeld, J. and Flexer, A. 1994. Response of benthic foraminifera to various pollution sources: implications for pollution monitoring. *Journal of Foraminiferal Research* 24, 1–17

Yassini, I. 1979. The littoral system ostracodes from the Bay of Bou-Ismail, Algeria. *Revista Española de Micropaleontologia* 11(3), 354–416

Zaghloul, Z.M., Abdel-Daiem, A.A. and Taha, A.A. 1990. Geomorphology, geologic evolution and subsidence of the Nile delta during the Quaternary. *Bulletin of the Faculty of Science*, Mansoura 17(1), 471–95

Index

The sunken portion of the ancient canopic region shown on a satellite picture of the west delta of the Nile.

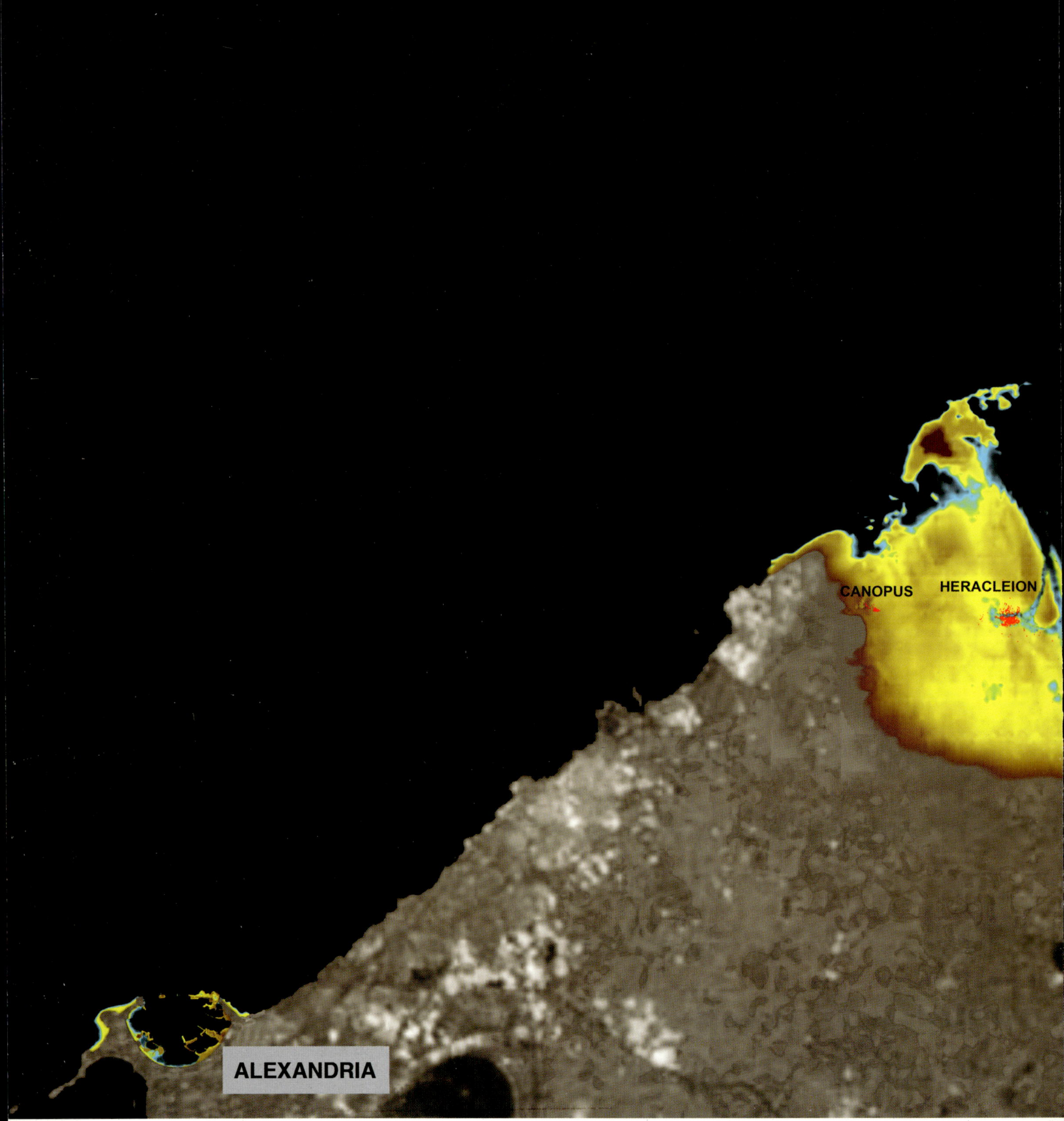